VOL. LII

ARCHITECTS | CONTRACTORS | ENGINEERS

GUIDE TO CONSTRUCTION COSTS 2021 EDITION

Architects Contractors Engineers DCD Guide to Construction Costs 2021, Volume 52.

ISBN 978-1-58855-192-4

EDITOR'S NOTE 2021

This annually published book is designed to give a uniform estimating and cost control system to the General Building Contractor. It contains a complete system to be used with or without computers. It also contains Quick Estimating sections for preliminary conceptual budget estimates by Architects, Engineers and Contractors. Square Foot Estimating is also included for preliminary estimates.

The Metropolitan Area concept is also used and gives the cost modifiers to use for the variations between Metropolitan Areas. This encompasses over 80% of the industry. This book is published annually to be historically accurate with the traditional May-July wage contract settlements and to be a true construction year estimating and cost guide.

The Rate of Inflation in the Construction Industry in 2020 was 2.5%. Labor contributed a 1.9% increase and materials 3%.

The Wage Rate for Skilled Trades increased an average of 1.9% in 2020. Wage rates will probably increase at a 1.5% average next year.

The Material Rate was up in 2020. There were several significant changes. Price increases are rare, with some noticeable decreases, such as Portland cement.

Construction Volume should be down for 2021. Housing and Commercial Construction spending will be down. Highway and Heavy Construction should be strong; there is still talk of improving our infrastructure and now an economic reason to spend.

The Construction Industry has had low to moderate inflation. Some materials will inflate or deflate, so watch them carefully.

We are recommending using a 2% increase in your estimates for work beyond July 1, 2020.

CONTENTS

Metro Area Multipliers

The costs as presented in this book attempt to represent national averages. Costs, however, vary among regions, states and even between adjacent localities.

In order to more closely approximate the probable costs for specific locations throughout the U.S., this table of modifiers is provided in the following pages. These adjustment factors are used to modify costs obtained from this book to help account for regional variations of construction costs and to provide a more accurate estimate for specific areas. The factors are formulated by comparing costs in a specific area to the costs as presented in the Costbook pages. An example of how to use these factors is shown below. Whenever local current costs are known, whether material prices or labor rates, they should be used when more accuracy is required.

Cost from Costbook Pages X Metro Area Multiplier Divided by 100 = Adjusted Cost

For example, a project estimated to cost $1,000,000 using the Costbook pages can be adjusted to more closely approximate the cost in Los Angeles, where the Multiplier is 133:

$$\$1,000,000 \quad X \quad \frac{133}{100} \quad = \quad \$1,330,000$$

1

State	Metropolitan Area	Multiplier
AK	ANCHORAGE	130
AL	ANNISTON	77
	AUBURN	75
	BIRMINGHAM	78
	DECATUR	75
	DOTHAN	75
	FLORENCE	75
	GADSDEN	77
	HUNTSVILLE	76
	MOBILE	76
	MONTGOMERY	76
	OPELIKA	75
	TUSCALOOSA	79
AR	FAYETTEVILLE	73
	FORT SMITH	73
	JONESBORO	74
	LITTLE ROCK	77
	NORTH LITTLE ROCK	77
	PINE BLUFF	75
	ROGERS	79
	SPRINGDALE	74
	TEXARKANA	75
AZ	FLAGSTAFF	80
	MESA	80
	PHOENIX	82
	TUCSON	76
	YUMA	76
CA	BAKERSFIELD	130
	CHICO	136
	FAIRFIELD	136
	FRESNO	139
	LODI	139
	LONG BEACH	133
	LOS ANGELES	133
	MERCED	139
	MODESTO	139
	NAPA	136
	OAKLAND	139
	ORANGE COUNTY	133
	PARADISE	136
	PORTERVILLE	132
	REDDING	136
	RIVERSIDE	133
	SACRAMENTO	136
	SALINAS	139
	SAN BERNARDINO	133
	SAN DIEGO	130
	SAN FRANCISCO	139

Metro Area Multipliers

State	Metropolitan Area	Multiplier
CA	SAN JOSE	139
	SAN LUIS OBISPO	133
	SANTA BARBARA	133
	SANTA CRUZ	139
	SANTA ROSA	136
	STOCKTON	139
	TULARE	132
	VALLEJO	136
	VENTURA	133
	VISALIA	132
	WATSONVILLE	139
	YOLO	136
	YUBA CITY	136
CO	BOULDER	90
	COLORADO SPRINGS	89
	DENVER	90
	FORT COLLINS	86
	GRAND JUNCTION	87
	GREELEY	88
	LONGMONT	88
	LOVELAND	90
	PUEBLO	87
CT	BRIDGEPORT	127
	DANBURY	127
	HARTFORD	122
	MERIDEN	125
	NEW HAVEN	125
	NEW LONDON	122
	NORWALK	127
	NORWICH	122
	STAMFORD	127
	WATERBURY	125
DC	WASHINGTON	102
DE	DOVER	123
	NEWARK	122
	WILMINGTON	114
FL	BOCA RATON	84
	BRADENTON	82
	CAPE CORAL	81
	CLEARWATER	83
	DAYTONA BEACH	81
	FORT LAUDERDALE	85
	FORT MYERS	79
	FORT PIERCE	81
	FORT WALTON BEACH	79
	GAINESVILLE	79
	JACKSONVILLE	79
	LAKELAND	79

Metro Area Multipliers		
State	**Metropolitan Area**	**Multiplier**
FL	MELBOURNE	81
	MIAMI	85
	NAPLES	82
	OCALA	80
	ORLANDO	81
	PALM BAY	82
	PANAMA CITY	79
	PENSACOLA	78
	PORT ST. LUCIE	81
	PUNTA GORDA	81
	SARASOTA	81
	ST. PETERSBURG	85
	TALLAHASSEE	80
	TAMPA	84
	TITUSVILLE	94
	WEST PALM BEACH	85
	WINTER HAVEN	85
GA	ALBANY	80
	ATHENS	80
	ATLANTA	77
	AUGUSTA	79
	COLUMBUS	76
	MACON	78
	SAVANNAH	79
HI	HONOLULU	133
IA	CEDAR FALLS	100
	CEDAR RAPIDS	105
	DAVENPORT	111
	DES MOINES	111
	DUBUQUE	104
	IOWA CITY	107
	SIOUX CITY	100
	WATERLOO	100
ID	BOISE CITY	83
	POCATELLO	88
IL	BLOOMINGTON	126
	CHAMPAIGN	121
	CHICAGO	144
	DECATUR	119
	KANKAKEE	126
	NORMAL	126
	PEKIN	123
	PEORIA	123
	ROCKFORD	126
	SPRINGFIELD	119
	URBANA	121
IN	BLOOMINGTON	105
	ELKHART	105

	Metro Area Multipliers	
State	**Metropolitan Area**	**Multiplier**
IN	EVANSVILLE	104
	FORT WAYNE	105
	GARY	119
	GOSHEN	105
	INDIANAPOLIS	105
	KOKOMO	105
	LAFAYETTE	105
	MUNCIE	105
	SOUTH BEND	119
	TERRE HAUTE	104
KS	KANSAS CITY	111
	LAWRENCE	109
	TOPEKA	108
	WICHITA	97
KY	LEXINGTON	100
	LOUISVILLE	100
	OWENSBORO	99
LA	ALEXANDRIA	80
	BATON ROUGE	82
	BOSSIER CITY	82
	HOUMA	81
	LAFAYETTE	81
	LAKE CHARLES	81
	MONROE	80
	NEW ORLEANS	81
	SHREVEPORT	82
MA	BARNSTABLE	139
	BOSTON	139
	BROCKTON	134
	FITCHBURG	133
	LAWRENCE	139
	LEOMINSTER	133
	LOWELL	139
	NEW BEDFORD	139
	PITTSFIELD	133
	SPRINGFIELD	125
	WORCESTER	133
	YARMOUTH	139
MD	BALTIMORE	89
	CUMBERLAND	91
	HAGERSTOWN	91
ME	AUBURN	84
	BANGOR	83
	LEWISTON	84
	PORTLAND	84
MI	ANN ARBOR	118
	BATTLE CREEK	103
	BAY CITY	105

Metro Area Multipliers		
State	**Metropolitan Area**	**Multiplier**
MI	BENTON HARBOR	118
	DETROIT	118
	EAST LANSING	108
	FLINT	110
	GRAND RAPIDS	87
	HOLLAND	93
	JACKSON	110
	KALAMAZOO	103
	LANSING	108
	MIDLAND	101
	MUSKEGON	93
	SAGINAW	103
MN	DULUTH	122
	MINNEAPOLIS	122
	ROCHESTER	122
	ST. CLOUD	118
	ST. PAUL	123
MO	COLUMBIA	114
	JOPLIN	97
	KANSAS CITY	115
	SPRINGFIELD	98
	ST. JOSEPH	114
	ST. LOUIS	115
MS	BILOXI	69
	GULFPORT	69
	HATTIESBURG	69
	JACKSON	69
	PASCAGOULA	69
MT	BILLINGS	98
	GREAT FALLS	96
	MISSOULA	98
NC	ASHEVILLE	74
	CHAPEL HILL	71
	CHARLOTTE	72
	DURHAM	71
	FAYETTEVILLE	71
	GOLDSBORO	71
	GREENSBORO	71
	GREENVILLE	71
	HICKORY	72
	HIGH POINT	72
	JACKSONVILLE	74
	LENOIR	74
	MORGANTON	72
	RALEIGH	70
	ROCKY MOUNT	72
	WILMINGTON	74
	WINSTON SALEM	71

Metro Area Multipliers		
State	**Metropolitan Area**	**Multiplier**
ND	BISMARCK	91
	FARGO	94
	GRAND FORKS	92
NE	LINCOLN	92
	OMAHA	92
NH	MANCHESTER	89
	NASHUA	89
	PORTSMOUTH	90
NJ	ATLANTIC CITY	140
	BERGEN	143
	BRIDGETON	138
	CAPE MAY	138
	HUNTERDON	141
	JERSEY CITY	143
	MIDDLESEX	142
	MILLVILLE	138
	MONMOUTH	127
	NEWARK	143
	OCEAN	138
	PASSAIC	144
	SOMERSET	141
	TRENTON	128
	VINELAND	138
NM	ALBUQUERQUE	81
	LAS CRUCES	81
	SANTA FE	80
NV	LAS VEGAS	128
	RENO	125
NY	ALBANY	116
	BINGHAMTON	113
	BUFFALO	111
	DUTCHESS COUNTY	116
	ELMIRA	114
	GLENS FALLS	114
	JAMESTOWN	110
	NASSAU	116
	NEW YORK	164
	NEWBURGH	116
	NIAGARA FALLS	125
	ROCHESTER	114
	ROME	116
	SCHENECTADY	116
	SUFFOLK	154
	SYRACUSE	112
	TROY	116
	UTICA	112
OH	AKRON	107
	CANTON	102

Metro Area Multipliers

State	Metropolitan Area	Multiplier
OH	CINCINNATI	103
	CLEVELAND	111
	COLUMBUS	104
	DAYTON	102
	ELYRIA	107
	HAMILTON	103
	LIMA	104
	LORAIN	107
	MANSFIELD	104
	MASSILLON	102
	MIDDLETOWN	102
	SPRINGFIELD	103
	STEUBENVILLE	104
	TOLEDO	107
	WARREN	107
	YOUNGSTOWN	107
OK	ENID	80
	LAWTON	83
	OKLAHOMA CITY	80
	TULSA	82
OR	ASHLAND	103
	CORVALLIS	110
	EUGENE	108
	MEDFORD	103
	PORTLAND	113
	SALEM	110
	SPRINGFIELD	108
PA	ALLENTOWN	118
	ALTOONA	112
	BETHLEHEM	120
	CARLISLE	111
	EASTON	120
	ERIE	110
	HARRISBURG	111
	HAZLETON	115
	JOHNSTOWN	111
	LANCASTER	107
	LEBANON	107
	PHILADELPHIA	133
	PITTSBURGH	110
	READING	119
	SCRANTON	114
	SHARON	110
	STATE COLLEGE	112
	WILKES BARRE	115
	WILLIAMSPORT	115
	YORK	109

Metro Area Multipliers		
State	**Metropolitan Area**	**Multiplier**
PR	MAYAGUEZ	72
	PONCE	72
	SAN JUAN	72
RI	PROVIDENCE	125
SC	AIKEN	71
	ANDERSON	69
	CHARLESTON	71
	COLUMBIA	71
	FLORENCE	72
	GREENVILLE	70
	MYRTLE BEACH	69
	NORTH CHARLESTON	71
	SPARTANBURG	71
	SUMTER	71
SD	RAPID CITY	78
	SIOUX FALLS	84
TN	CHATTANOOGA	78
	CLARKSVILLE	77
	JACKSON	76
	JOHNSON CITY	78
	KNOXVILLE	75
	MEMPHIS	78
	NASHVILLE	77
TX	ABILENE	74
	AMARILLO	74
	ARLINGTON	73
	AUSTIN	76
	BEAUMONT	76
	BRAZORIA	76
	BROWNSVILLE	70
	BRYAN	75
	COLLEGE STATION	75
	CORPUS CHRISTI	74
	DALLAS	74
	DENISON	74
	EDINBURG	70
	EL PASO	73
	FORT WORTH	73
	GALVESTON	75
	HARLINGEN	70
	HOUSTON	71
	KILLEEN	73
	LAREDO	72
	LONGVIEW	73
	LUBBOCK	75
	MARSHALL	69
	MCALLEN	70

State	Metropolitan Area	Multiplier
	Metro Area Multipliers	
TX	MIDLAND	73
	MISSION	70
	ODESSA	73
	PORT ARTHUR	76
	SAN ANGELO	73
	SAN ANTONIO	77
	SAN BENITO	70
	SAN MARCOS	75
	SHERMAN	70
	TEMPLE	73
	TEXARKANA	73
	TEXAS CITY	70
	TYLER	72
	VICTORIA	74
	WACO	73
	WICHITA FALLS	75
UT	OGDEN	77
	OREM	76
	PROVO	76
	SALT LAKE CITY	77
VA	CHARLOTTESVILLE	80
	LYNCHBURG	81
	NEWPORT NEWS	82
	NORFOLK	82
	PETERSBURG	79
	RICHMOND	80
	ROANOKE	83
	VIRGINIA BEACH	82
VT	BURLINGTON	80
WA	BELLEVUE	116
	BELLINGHAM	109
	BREMERTON	111
	EVERETT	115
	KENNEWICK	112
	OLYMPIA	114
	PASCO	112
	RICHLAND	112
	SEATTLE	116
	SPOKANE	93
	TACOMA	116
	YAKIMA	102
WI	APPLETON	112
	BELOIT	115
	EAU CLAIRE	112
	GREEN BAY	111
	JANESVILLE	115
	KENOSHA	116
	LA CROSSE	112

Metro Area Multipliers		
State	Metropolitan Area	Multiplier
WI	MADISON	114
	MILWAUKEE	118
	NEENAH	112
	OSHKOSH	112
	RACINE	117
	SHEBOYGAN	111
	WAUKESHA	118
	WAUSAU	111
WV	CHARLESTON	114
	HUNTINGTON	117
	PARKERSBURG	112
	WHEELING	110
WY	CASPER	85
	CHEYENNE	86

TABLE OF CONTENTS	PAGE

HOW TO USE THIS BOOK

Labor Columns

> ➤ Units *include* Workers Comp., Unemployment, and FICA on labor (approximately 35%).
> ➤ Units *do not include* general conditions and equipment (approximately 10%).
> ➤ Units *do not include* contractors' overhead and profit (approximately 10%).
> ➤ Units are Government Prevailing wages.

Material Columns

> ➤ Units *do not include* general conditions and equipment (approximately 10%).
> ➤ Units *do not include* sales or use taxes (approximately 5%) of material cost.
> ➤ Units *do not include* contractors' overhead and profit (approximately 10%) and are FOB job site.

Total Columns (Subcontractors)

> ➤ Units *do not include* general contractors' overhead or profit (approximately 10%).

Quick Estimating Sections - For Preliminary and Conceptual Estimating

> ➤ Includes all labor, material, general conditions, equipment, taxes.

		UNIT	LABOR	MAT.	TOTAL
01020.10	**ALLOWANCES**				
0090	Overhead				
1000	$20,000 project				
1040	Average	PCT			20.00
1080	$100,000 project				
1120	Average	PCT			15.00
1160	$500,000 project				
1180	Average	PCT			12.00
1220	$1,000,000 project				
1260	Average	PCT			10.00
1480	Profit				
1500	$20,000 project				
1540	Average	PCT			15.00
1580	$100,000 project				
1620	Average	PCT			12.00
1660	$500,000 project				
1700	Average	PCT			10.00
1740	$1,000,000 project				
1780	Average	PCT			8.00
2000	Professional fees				
2100	Architectural				
2120	$100,000 project				
2160	Average	PCT			10.00
2200	$500,000 project				
2240	Average	PCT			8.00
2280	$1,000,000 project				
2320	Average	PCT			7.00
4080	Taxes				
5000	Sales tax				
5040	Average	PCT			5.00
5080	Unemployment				
5120	Average	PCT			6.50
5200	Social security (FICA)	"			7.85
01050.10	**FIELD STAFF**				
1000	Superintendent				
1020	Minimum	YEAR			102,000
1040	Average	"			127,000
1060	Maximum	"			153,000
1080	Field engineer				
1100	Minimum	YEAR			100,000
1120	Average	"			115,000
1140	Maximum	"			132,000
1160	Foreman				
1180	Minimum	YEAR			68,000
1200	Average	"			108,000
1220	Maximum	"			127,000
1240	Bookkeeper/timekeeper				
1260	Minimum	YEAR			39,200
1280	Average	"			51,300
1300	Maximum	"			66,300
1320	Watchman				
1340	Minimum	YEAR			29,300

01050.10	FIELD STAFF, Cont'd...	UNIT	LABOR	MAT.	TOTAL
1360	Average	YEAR			39,000
1380	Maximum	"			49,400
01310.10	**SCHEDULING**				
0090	Scheduling for				
1000	$100,000 project				
1040	Average	PCT			2.05
1080	$500,000 project				
1120	Average	PCT			1.02
1160	$1,000,000 project				
1200	Average	PCT			0.77
4000	Scheduling software				
4020	Minimum	EA			690
4040	Average	"			3,970
4060	Maximum	"			79,410
01410.10	**TESTING**				
1080	Testing concrete, per test				
1100	Minimum	EA			23.50
1120	Average	"			39.25
1140	Maximum	"			78.00
1160	Soil, per test				
1180	Minimum	EA			48.00
1200	Average	"			120
1220	Maximum	"			320
01500.10	**TEMPORARY FACILITIES**				
1000	Barricades, temporary				
1010	Highway				
1020	Concrete	LF	5.10	14.75	19.85
1040	Wood	"	2.04	5.10	7.14
1060	Steel	"	1.70	5.29	6.99
1080	Pedestrian barricades				
1100	Plywood	SF	1.70	4.54	6.24
1120	Chain link fence	"	1.70	3.85	5.55
2000	Trailers, general office type, per month				
2020	Minimum	EA			250
2040	Average	"			410
2060	Maximum	"			820
2080	Crew change trailers, per month				
2100	Minimum	EA			150
2120	Average	"			160
2140	Maximum	"			250
01505.10	**MOBILIZATION**				
1000	Equipment mobilization				
1020	Bulldozer				
1040	Minimum	EA			220
1060	Average	"			460
1080	Maximum	"			780
1100	Backhoe/front-end loader				
1120	Minimum	EA			130
1140	Average	"			230
1160	Maximum	"			510
1180	Crane, crawler type				

		UNIT	LABOR	MAT.	TOTAL
01505.10	**MOBILIZATION, Cont'd...**				
1200	Minimum	EA			2,440
1220	Average	"			6,000
1240	Maximum	"			12,880
1260	Truck crane				
1280	Minimum	EA			560
1300	Average	"			860
1320	Maximum	"			1,490
1340	Pile driving rig				
1360	Minimum	EA			11,100
1380	Average	"			22,210
1400	Maximum	"			39,970
01525.10	**CONSTRUCTION AIDS**				
1000	Scaffolding/staging, rent per month				
1020	Measured by lineal feet of base				
1040	10' high	LF			14.50
1060	20' high	"			26.50
1080	30' high	"			37.00
1100	40' high	"			42.75
1120	50' high	"			51.00
1140	Measured by square foot of surface				
1160	Minimum	SF			0.64
1180	Average	"			1.11
1200	Maximum	"			1.99
1220	Safety nets, heavy duty, per job				
1240	Minimum	SF			0.43
1260	Average	"			0.52
1280	Maximum	"			1.14
1300	Tarpaulins, fabric, per job				
1320	Minimum	SF			0.30
1340	Average	"			0.51
1360	Maximum	"			1.32
01525.20	**TEMPORARY CONST. SHELTERS**				
0010	Standard, alum. with fabric, 12'x20'x15' Ht.	SF			20.00
0020	12'x20'x20' Ht.	"			23.00
01570.10	**SIGNS**				
0080	Construction signs, temporary				
1000	Signs, 2' x 4'				
1020	Minimum	EA			41.75
1040	Average	"			100
1060	Maximum	"			350
1160	Signs, 8' x 8'				
1180	Minimum	EA			110
1200	Average	"			350
1220	Maximum	"			3,540
01600.10	**EQUIPMENT**				
0080	Air compressor				
1000	60 cfm				
1020	By day	EA			110
1030	By week	"			320
1040	By month	"			980
1200	600 cfm				

01600.10	EQUIPMENT, Cont'd...	UNIT	LABOR	MAT.	TOTAL
1210	By day	EA			620
1220	By week	"			1,860
1230	By month	"			5,630
1300	Air tools, per compressor, per day				
1310	Minimum	EA			44.50
1320	Average	"			56.00
1330	Maximum	"			78.00
1400	Generators, 5 kw				
1410	By day	EA			110
1420	By week	"			330
1430	By month	"			1,020
1500	Heaters, salamander type, per week				
1510	Minimum	EA			130
1520	Average	"			190
1530	Maximum	"			400
1600	Pumps, submersible				
1605	50 gpm				
1610	By day	EA			89.00
1620	By week	"			270
1630	By month	"			800
1675	500 gpm				
1680	By day	EA			180
1690	By week	"			530
1700	By month	"			1,600
1900	Diaphragm pump, by week				
1920	Minimum	EA			160
1930	Average	"			270
1940	Maximum	"			550
2000	Pickup truck				
2020	By day	EA			170
2030	By week	"			490
2040	By month	"			1,510
2080	Dump truck				
2100	6 c.y. truck				
2120	By day	EA			440
2130	By week	"			1,330
2140	By month	"			4,010
2300	16 c.y. truck				
2310	By day	EA			890
2320	By week	"			2,670
2340	By month	"			8,020
2400	Backhoe, track mounted				
2420	1/2 c.y. capacity				
2430	By day	EA			910
2440	By week	"			2,780
2450	By month	"			8,240
2500	1 c.y. capacity				
2510	By day	EA			1,450
2520	By week	"			4,340
2530	By month	"			13,030
2600	3 c.y. capacity				
2620	By day	EA			4,680

		UNIT	LABOR	MAT.	TOTAL
01600.10	**EQUIPMENT, Cont'd...**				
2640	By week	EA			14,030
2680	By month	"			42,100
3000	Backhoe/loader, rubber tired				
3005	1/2 c.y. capacity				
3010	By day	EA			550
3020	By week	"			1,670
3030	By month	"			5,010
3035	3/4 c.y. capacity				
3040	By day	EA			670
3050	By week	"			2,000
3060	By month	"			6,010
3200	Bulldozer				
3205	75 hp				
3210	By day	EA			780
3220	By week	"			2,340
3230	By month	"			7,020
4000	Cranes, crawler type				
4005	15 ton capacity				
4010	By day	EA			1,000
4020	By week	"			3,010
4030	By month	"			9,020
4070	50 ton capacity				
4080	By day	EA			2,230
4090	By week	"			6,680
4100	By month	"			20,050
4145	Truck mounted, hydraulic				
4150	15 ton capacity				
4160	By day	EA			940
4170	By week	"			2,840
4180	By month	"			8,190
5380	Loader, rubber tired				
5385	1 c.y. capacity				
5390	By day	EA			670
5400	By week	"			2,000
5410	By month	"			6,020
7000	Scraper				
7010	Elevated scraper, not including bulldozer, 12 c.y.				
7020	By day	EA			1,520
7030	By week	"			4,570
7040	By month	"			12,990
7100	Self-propelled scraper, 14 c.y.				
7110	By day	EA			3,060
7120	By week	"			9,170
7130	By month	"			25,890
01740.10	**BONDS**				
1000	Performance bonds				
1020	Minimum	PCT			0.62
1040	Average	"			1.93
1060	Maximum	"			3.07

TABLE OF CONTENTS   PAGE

		UNIT	LABOR	MAT.	TOTAL
02210.10	**SOIL BORING**				
1000	Borings, uncased, stable earth				
1020	2-1/2" dia.	LF	35.75		35.75
1040	4" dia.	"	40.75		40.75
1500	Cased, including samples				
1520	2-1/2" dia.	LF	47.50		47.50
1540	4" dia.	"	81.00		81.00
02220.10	**COMPLETE BUILDING DEMOLITION**				
0200	Wood frame	CF	0.41		0.41
0300	Concrete	"	0.61		0.61
0400	Steel frame	"	0.82		0.82
02220.15	**SELECTIVE BUILDING DEMOLITION**				
1000	Partition removal				
1100	Concrete block partitions				
1120	4" thick	SF	2.55		2.55
1140	8" thick	"	3.40		3.40
1160	12" thick	"	4.64		4.64
1200	Brick masonry partitions				
1220	4" thick	SF	2.55		2.55
1240	8" thick	"	3.19		3.19
1260	12" thick	"	4.25		4.25
1280	16" thick	"	6.38		6.38
1380	Cast-in-place concrete partitions				
1400	Unreinforced				
1421	6" thick	SF	19.00		19.00
1423	8" thick	"	20.25		20.25
1425	10" thick	"	23.75		23.75
1427	12" thick	"	28.50		28.50
1440	Reinforced				
1441	6" thick	SF	22.00		22.00
1443	8" thick	"	28.50		28.50
1445	10" thick	"	31.75		31.75
1447	12" thick	"	38.00		38.00
1500	Terra cotta				
1520	To 6" thick	SF	2.55		2.55
1700	Stud partitions				
1720	Metal or wood, with drywall both sides	SF	2.55		2.55
1740	Metal studs, both sides, lath and plaster	"	3.40		3.40
2000	Door and frame removal				
2020	Hollow metal in masonry wall				
2030	Single				
2040	2'6"x6'8"	EA	64.00		64.00
2060	3'x7'	"	85.00		85.00
2070	Double				
2080	3'x7'	EA	100		100
2085	4'x8'	"	100		100
2140	Wood in framed wall				
2150	Single				
2160	2'6"x6'8"	EA	36.50		36.50
2180	3'x6'8"	"	42.50		42.50
2190	Double				
2200	2'6"x6'8"	EA	51.00		51.00

		UNIT	LABOR	MAT.	TOTAL
02220.15	**SELECTIVE BUILDING DEMOLITION, Cont'd...**				
2220	3'x6'8"	EA	57.00		57.00
2240	Remove for re-use				
2260	Hollow metal	EA	130		130
2280	Wood	"	85.00		85.00
2300	Floor removal				
2340	Brick flooring	SF	2.04		2.04
2360	Ceramic or quarry tile	"	1.13		1.13
2380	Terrazzo	"	2.27		2.27
2400	Heavy wood	"	1.36		1.36
2420	Residential wood	"	1.45		1.45
2440	Resilient tile or linoleum	"	0.51		0.51
2500	Ceiling removal				
2520	Acoustical tile ceiling				
2540	Adhesive fastened	SF	0.51		0.51
2560	Furred and glued	"	0.42		0.42
2580	Suspended grid	"	0.31		0.31
2600	Drywall ceiling				
2620	Furred and nailed	SF	0.56		0.56
2640	Nailed to framing	"	0.51		0.51
2660	Plastered ceiling				
2680	Furred on framing	SF	1.27		1.27
2700	Suspended system	"	1.70		1.70
2800	Roofing removal				
2820	Steel frame				
2840	Corrugated metal roofing	SF	1.02		1.02
2860	Built-up roof on metal deck	"	1.70		1.70
2900	Wood frame				
2920	Built-up roof on wood deck	SF	1.57		1.57
2940	Roof shingles	"	0.85		0.85
2960	Roof tiles	"	1.70		1.70
8900	Concrete frame	CF	3.40		3.40
8920	Concrete plank	SF	2.55		2.55
8940	Built-up roof on concrete	"	1.45		1.45
9200	Cut-outs				
9230	Concrete, elevated slabs, mesh reinforcing				
9240	Under 5 cf	CF	51.00		51.00
9260	Over 5 cf	"	42.50		42.50
9270	Bar reinforcing				
9280	Under 5 cf	CF	85.00		85.00
9290	Over 5 cf	"	64.00		64.00
9300	Window removal				
9301	Metal windows, trim included				
9302	2'x3'	EA	51.00		51.00
9304	2'x4'	"	57.00		57.00
9306	2'x6'	"	64.00		64.00
9308	3'x4'	"	64.00		64.00
9310	3'x6'	"	73.00		73.00
9312	3'x8'	"	85.00		85.00
9314	4'x4'	"	85.00		85.00
9315	4'x6'	"	100		100
9316	4'x8'	"	130		130
9317	Wood windows, trim included				

		UNIT	LABOR	MAT.	TOTAL
02220.15	**SELECTIVE BUILDING DEMOLITION, Cont'd...**				
9318	2'x3'	EA	28.25		28.25
9319	2'x4'	"	30.00		30.00
9320	2'x6'	"	32.00		32.00
9321	3'x4'	"	34.00		34.00
9322	3'x6'	"	36.50		36.50
9324	3'x8'	"	39.25		39.25
9325	6'x4'	"	42.50		42.50
9326	6'x6'	"	46.50		46.50
9327	6'x8'	"	51.00		51.00
9329	Walls, concrete, bar reinforcing				
9330	Small jobs	CF	34.00		34.00
9340	Large jobs	"	28.25		28.25
9360	Brick walls, not including toothing				
9390	4" thick	SF	2.55		2.55
9400	8" thick	"	3.19		3.19
9410	12" thick	"	4.25		4.25
9415	16" thick	"	6.38		6.38
9420	Concrete block walls, not including toothing				
9440	4" thick	SF	2.83		2.83
9450	6" thick	"	3.00		3.00
9460	8" thick	"	3.19		3.19
9465	10" thick	"	3.64		3.64
9470	12" thick	"	4.25		4.25
9500	Rubbish handling				
9519	Load in dumpster or truck				
9520	Minimum	CF	1.13		1.13
9540	Maximum	"	1.70		1.70
9550	For use of elevators, add				
9560	Minimum	CF	0.25		0.25
9570	Maximum	"	0.51		0.51
9600	Rubbish hauling				
9640	Hand loaded on trucks, 2 mile trip	CY	44.25		44.25
9660	Machine loaded on trucks, 2 mile trip	"	28.50		28.50
02225.20	**FENCE DEMOLITION**				
0060	Remove fencing				
0080	Chain link, 8' high				
0100	For disposal	LF	2.55		2.55
0200	For reuse	"	6.38		6.38
0980	Wood				
1000	4' high	SF	1.70		1.70
1960	Masonry				
1980	8" thick				
2000	4' high	SF	5.10		5.10
2020	6' high	"	6.38		6.38
02225.50	**SAW CUTTING PAVEMENT**				
0100	Pavement, bituminous				
0110	2" thick	LF	2.21		2.21
0120	3" thick	"	2.76		2.76
0200	Concrete pavement, with wire mesh				
0210	4" thick	LF	4.25		4.25
0212	5" thick	"	4.61		4.61

		UNIT	LABOR	MAT.	TOTAL
02225.50	**SAW CUTTING PAVEMENT, Cont'd...**				
0300	Plain concrete, unreinforced				
0320	4" thick	LF	3.68		3.68
0340	5" thick	"	4.25		4.25
02225.60	**TORCH CUTTING**				
0010	Steel plate, 1/2" thick	LF	1.13		1.13
0020	1" thick	"	2.26		2.26
02230.50	**TREE CUTTING & CLEARING**				
0980	Cut trees and clear out stumps				
1000	9" to 12" dia.	EA	570		570
1400	To 24" dia.	"	710		710
1600	24" dia. and up	"	950		950
02315.10	**BASE COURSE**				
1019	Base course, crushed stone				
1020	3" thick	SY	0.77	3.19	3.96
1030	4" thick	"	0.84	4.29	5.13
1040	6" thick	"	0.91	6.43	7.34
2500	Base course, bank run gravel				
3020	4" deep	SY	0.81	3.02	3.83
3040	6" deep	"	0.88	4.62	5.50
4000	Prepare and roll sub-base				
4020	Minimum	SY	0.77		0.77
4030	Average	"	0.97		0.97
4040	Maximum	"	1.29		1.29
02315.20	**BORROW**				
1000	Borrow fill, FOB at pit				
1005	Sand, haul to site, round trip				
1010	10 mile	CY	15.50	22.75	38.25
1020	20 mile	"	26.00	22.75	48.75
1030	30 mile	"	38.75	22.75	61.50
3980	Place borrow fill and compact				
4000	Less than 1 in 4 slope	CY	7.77	22.75	30.52
4100	Greater than 1 in 4 slope	"	10.25	22.75	33.00
02315.30	**BULK EXCAVATION**				
1000	Excavation, by small dozer				
1020	Large areas	CY	2.21		2.21
1040	Small areas	"	3.68		3.68
1060	Trim banks	"	5.53		5.53
1700	Hydraulic excavator				
1720	1 c.y. capacity				
1740	Light material	CY	4.75		4.75
1760	Medium material	"	5.70		5.70
1780	Wet material	"	7.12		7.12
1790	Blasted rock	"	8.14		8.14
1800	1-1/2 c.y. capacity				
1820	Light material	CY	1.94		1.94
1840	Medium material	"	2.59		2.59
1860	Wet material	"	3.10		3.10
2000	Wheel mounted front-end loader				
2020	7/8 c.y. capacity				
2040	Light material	CY	3.88		3.88

		UNIT	LABOR	MAT.	TOTAL
02315.30	**BULK EXCAVATION, Cont'd...**				
2060	Medium material	CY	4.44		4.44
2080	Wet material	"	5.18		5.18
2100	Blasted rock	"	6.21		6.21
2300	2-1/2 c.y. capacity				
2320	Light material	CY	1.82		1.82
2340	Medium material	"	1.94		1.94
2360	Wet material	"	2.07		2.07
2380	Blasted rock	"	2.22		2.22
2600	Track mounted front-end loader				
2620	1-1/2 c.y. capacity				
2640	Light material	CY	2.59		2.59
2660	Medium material	"	2.82		2.82
2680	Wet material	"	3.10		3.10
2700	Blasted rock	"	3.45		3.45
2720	2-3/4 c.y. capacity				
2740	Light material	CY	1.55		1.55
2760	Medium material	"	1.72		1.72
2780	Wet material	"	1.94		1.94
2790	Blasted rock	"	2.22		2.22
4000	Scraper - 500' haul				
4010	Elevated scraper, not including bulldozer, 12 c.y.				
4020	Light material	CY	5.18		5.18
4030	Medium material	"	5.65		5.65
4040	Wet material	"	6.21		6.21
4050	Blasted rock	"	6.91		6.91
4100	Self-propelled scraper, 14 c.y.				
4110	Light material	CY	4.78		4.78
4120	Medium material	"	5.18		5.18
4130	Wet material	"	5.65		5.65
4140	Blasted rock	"	6.21		6.21
4200	1,000' haul				
4210	Elevated scraper, not including bulldozer, 12 c.y.				
4220	Light material	CY	6.21		6.21
4230	Medium material	"	6.91		6.91
4240	Wet material	"	7.77		7.77
4250	Blasted rock	"	8.88		8.88
4300	Self-propelled scraper, 14 c.y.				
4310	Light material	CY	5.65		5.65
4320	Medium material	"	6.21		6.21
4330	Wet material	"	6.91		6.91
4340	Blasted rock	"	7.77		7.77
4400	2,000' haul				
4410	Elevated scraper, not including bulldozer, 12 c.y.				
4420	Light material	CY	7.77		7.77
4430	Medium material	"	8.88		8.88
4440	Wet material	"	10.25		10.25
4450	Blasted rock	"	12.50		12.50
4510	Self-propelled scraper, 14 c.y.				
4520	Light material	CY	6.91		6.91
4530	Medium material	"	7.77		7.77
4540	Wet material	"	8.88		8.88
4550	Blasted rock	"	10.25		10.25

		UNIT	LABOR	MAT.	TOTAL
02315.40	**BUILDING EXCAVATION**				
0090	Structural excavation, unclassified earth				
0100	3/8 c.y. backhoe	CY	20.75		20.75
0110	3/4 c.y. backhoe	"	15.50		15.50
0120	1 c.y. backhoe	"	13.00		13.00
0600	Foundation backfill and compaction by machine	"	31.00		31.00
02315.45	**HAND EXCAVATION**				
0980	Excavation				
1000	To 2' deep				
1020	Normal soil	CY	57.00		57.00
1040	Sand and gravel	"	51.00		51.00
1060	Medium clay	"	64.00		64.00
1080	Heavy clay	"	73.00		73.00
1100	Loose rock	"	85.00		85.00
1200	To 6' deep				
1220	Normal soil	CY	73.00		73.00
1240	Sand and gravel	"	64.00		64.00
1260	Medium clay	"	85.00		85.00
1280	Heavy clay	"	100		100
1300	Loose rock	"	130		130
2020	Backfilling foundation without compaction, 6" lifts	"	32.00		32.00
2200	Compaction of backfill around structures or in trench				
2220	By hand with air tamper	CY	36.50		36.50
2240	By hand with vibrating plate tamper	"	34.00		34.00
2250	1 ton roller	"	55.00		55.00
5400	Miscellaneous hand labor				
5440	Trim slopes, sides of excavation	SF	0.08		0.08
5450	Trim bottom of excavation	"	0.10		0.10
5460	Excavation around obstructions and services	CY	170		170
02315.50	**ROADWAY EXCAVATION**				
0100	Roadway excavation				
0110	1/4 mile haul	CY	3.10		3.10
0120	2 mile haul	"	5.18		5.18
0130	5 mile haul	"	7.77		7.77
3000	Spread base course	"	3.88		3.88
3100	Roll and compact	"	5.18		5.18
02315.60	**TRENCHING**				
0100	Trenching and continuous footing excavation				
0980	By gradall				
1000	1 c.y. capacity				
1040	Medium soil	CY	4.78		4.78
1080	Loose rock	"	5.65		5.65
1090	Blasted rock	"	5.98		5.98
1095	By hydraulic excavator				
1100	1/2 c.y. capacity				
1140	Medium soil	CY	5.65		5.65
1180	Loose rock	"	6.91		6.91
1190	Blasted rock	"	7.77		7.77
1200	1 c.y. capacity				
1240	Medium soil	CY	3.88		3.88
1280	Loose rock	"	4.44		4.44
1300	Blasted rock	"	4.78		4.78

		UNIT	LABOR	MAT.	TOTAL
02315.60	**TRENCHING, Cont'd...**				
1600	2 c.y. capacity				
1640	Medium soil	CY	3.27		3.27
1680	Loose rock	"	3.65		3.65
1690	Blasted rock	"	3.88		3.88
3000	Hand excavation				
3100	Bulk, wheeled 100'				
3120	Normal soil	CY	57.00		57.00
3140	Sand or gravel	"	51.00		51.00
3160	Medium clay	"	73.00		73.00
3180	Heavy clay	"	100		100
3200	Loose rock	"	130		130
3300	Trenches, up to 2' deep				
3320	Normal soil	CY	64.00		64.00
3340	Sand or gravel	"	57.00		57.00
3360	Medium clay	"	85.00		85.00
3380	Heavy clay	"	130		130
3390	Loose rock	"	170		170
3400	Trenches, to 6' deep				
3420	Normal soil	CY	73.00		73.00
3440	Sand or gravel	"	64.00		64.00
3460	Medium clay	"	100		100
3480	Heavy clay	"	170		170
3500	Loose rock	"	260		260
3590	Backfill trenches				
3600	With compaction				
3620	By hand	CY	42.50		42.50
3640	By 60 hp tracked dozer	"	2.76		2.76
02315.70	**UTILITY EXCAVATION**				
2080	Trencher, sandy clay, 8" wide trench				
2100	18" deep	LF	2.45		2.45
2200	24" deep	"	2.76		2.76
2300	36" deep	"	3.16		3.16
6080	Trench backfill, 95% compaction				
7000	Tamp by hand	CY	32.00		32.00
7050	Vibratory compaction	"	25.50		25.50
7060	Trench backfilling, with borrow sand, place & compact	"	25.50	22.75	48.25
02315.80	**HAULING MATERIAL**				
0090	Haul material by 10 c.y. dump truck, round trip distance				
0100	1 mile	CY	6.14		6.14
0110	2 mile	"	7.37		7.37
0120	5 mile	"	10.00		10.00
0130	10 mile	"	11.00		11.00
0140	20 mile	"	12.25		12.25
0150	30 mile	"	14.75		14.75
02340.05	**SOIL STABILIZATION**				
0100	Straw bale secured with rebar	LF	1.70	7.54	9.24
0120	Filter barrier, 18" high filter fabric	"	5.10	1.82	6.92
0130	Sediment fence, 36" fabric with 6" mesh	"	6.38	4.32	10.70
1000	Soil stabilization with tar paper, burlap, straw and stakes	SF	0.07	0.36	0.43

DIVISION # 02 SITE CONSTRUCTION

		UNIT	LABOR	MAT.	TOTAL
02360.20	**SOIL TREATMENT**				
1100	Soil treatment, termite control pretreatment				
1120	Under slabs	SF	0.28	0.38	0.66
1140	By walls	"	0.34	0.38	0.72
02370.40	**RIPRAP**				
0100	Riprap				
0110	Crushed stone blanket, max size 2-1/2"	TON	82.00	35.25	117
0120	Stone, quarry run, 300 lb. stones	"	76.00	44.25	120
0130	400 lb. stones	"	71.00	46.00	117
0140	500 lb. stones	"	66.00	48.00	114
0150	750 lb. stones	"	62.00	49.75	112
0160	Dry concrete riprap in bags 3" thick, 80 lb. per bag	BAG	4.11	5.96	10.07
02455.60	**STEEL PILES**				
1000	H-section piles				
1010	8x8				
1020	36 lb/ft				
1021	30' long	LF	14.25	21.75	36.00
1022	40' long	"	11.25	21.75	33.00
5000	Tapered friction piles, fluted casing, up to 50'				
5002	With 4000 psi concrete no reinforcing				
5040	12" dia.	LF	8.51	21.50	30.01
5060	14" dia.	"	8.73	24.75	33.48
02455.65	**STEEL PIPE PILES**				
1000	Concrete filled, 3000# concrete, up to 40'				
1100	8" dia.	LF	12.25	24.00	36.25
1120	10" dia.	"	12.50	31.00	43.50
1140	12" dia.	"	13.00	36.00	49.00
2000	Pipe piles, non-filled				
2020	8" dia.	LF	9.46	21.75	31.21
2040	10" dia.	"	9.73	27.25	36.98
2060	12" dia.	"	10.00	33.25	43.25
2520	Splice				
2540	8" dia.	EA	100	97.00	197
2560	10" dia.	"	100	110	210
2580	12" dia.	"	130	120	250
2680	Standard point				
2700	8" dia.	EA	100	130	230
2740	10" dia.	"	100	170	270
2760	12" dia.	"	130	180	310
2880	Heavy duty point				
2900	8" dia.	EA	130	230	360
2920	10" dia.	"	130	320	450
2940	12" dia.	"	170	340	510
02455.80	**WOOD AND TIMBER PILES**				
0080	Treated wood piles, 12" butt, 8" tip				
0100	25' long	LF	17.00	18.00	35.00
0110	30' long	"	14.25	19.25	33.50
0120	35' long	"	12.25	19.25	31.50
0125	40' long	"	10.75	19.25	30.00

		UNIT	LABOR	MAT.	TOTAL
02465.50	**PRESTRESSED PILING**				
0980	Prestressed concrete piling, less than 60' long	LF	7.09	20.75	27.84
1000	10" sq.	"	7.40	29.00	36.40
1002	12" sq.				
1480	Straight cylinder, less than 60' long	LF	7.74	27.00	34.74
1500	12" dia.	"	7.92	36.50	44.42
1540	14" dia.				
02510.10	**WELLS**				
0980	Domestic water, drilled and cased	LF	85.00	30.50	116
1000	4" dia.	"	95.00	33.50	129
1020	6" dia.				
02510.40	**DUCTILE IRON PIPE**				
0990	Ductile iron pipe, cement lined, slip-on joints	LF	7.91	18.25	26.16
1000	4"	"	8.38	21.25	29.63
1010	6"	"	8.90	27.75	36.65
1020	8"				
1190	Mechanical joint pipe	LF	11.00	19.75	30.75
1200	4"	"	11.75	23.50	35.25
1210	6"	"	13.00	31.00	44.00
1220	8"				
1480	Fittings, mechanical joint				
1500	90 degree elbow	EA	34.00	230	264
1520	4"	"	39.25	300	339
1540	6"	"	51.00	430	481
1560	8"				
1700	45 degree elbow	EA	34.00	200	234
1720	4"	"	39.25	270	309
1740	6"	"	51.00	380	431
1760	8"				
02510.60	**PLASTIC PIPE**				
0110	PVC, class 150 pipe	LF	7.12	5.31	12.43
0120	4" dia.	"	7.70	10.00	17.70
0130	6" dia.	"	8.14	16.00	24.14
0140	8" dia.				
0165	Schedule 40 pipe	LF	3.00	1.34	4.34
0170	1-1/2" dia.	"	3.19	1.99	5.18
0180	2" dia.	"	3.40	3.01	6.41
0185	2-1/2" dia.	"	3.64	4.09	7.73
0190	3" dia.	"	4.25	5.78	10.03
0200	4" dia.	"	5.10	11.00	16.10
0210	6" dia.				
0240	90 degree elbows	EA	8.51	1.12	9.63
0250	1"	"	8.51	2.14	10.65
0260	1-1/2"	"	9.28	3.35	12.63
0270	2"	"	10.25	10.25	20.50
0280	2-1/2"	"	11.25	12.25	23.50
0290	3"	"	12.75	19.75	32.50
0300	4"	"	17.00	62.00	79.00
0310	6"				
0500	Couplings	EA	8.51	0.91	9.42
0510	1"	"	8.51	1.30	9.81
0520	1-1/2"				

		UNIT	LABOR	MAT.	TOTAL
02510.60	**PLASTIC PIPE, Cont'd...**				
0530	2"	EA	9.28	2.01	11.29
0540	2-1/2"	"	10.25	4.42	14.67
0550	3"	"	11.25	6.91	18.16
0560	4"	"	12.75	9.02	21.77
0580	6"	"	17.00	28.50	45.50
02530.20	**VITRIFIED CLAY PIPE**				
0100	Vitrified clay pipe, extra strength				
1020	6" dia.	LF	13.00	5.73	18.73
1040	8" dia.	"	13.50	6.87	20.37
1050	10" dia.	"	14.25	10.50	24.75
02530.30	**MANHOLES**				
0100	Precast sections, 48" dia.				
0110	Base section	EA	240	360	600
0120	1'0" riser	"	190	100	290
0130	1'4" riser	"	200	120	320
0140	2'8" riser	"	220	180	400
0150	4'0" riser	"	240	340	580
0160	2'8" cone top	"	290	220	510
0170	Precast manholes, 48" dia.				
0180	4' deep	EA	570	700	1,270
0200	6' deep	"	710	1,070	1,780
0250	7' deep	"	810	1,220	2,030
0260	8' deep	"	950	1,380	2,330
0280	10' deep	"	1,140	1,540	2,680
1000	Cast-in-place, 48" dia., with frame and cover				
1100	5' deep	EA	1,430	630	2,060
1120	6' deep	"	1,630	830	2,460
1140	8' deep	"	1,900	1,210	3,110
1160	10' deep	"	2,280	1,410	3,690
1480	Brick manholes, 48" dia. with cover, 8" thick				
1500	4' deep	EA	620	670	1,290
1501	6' deep	"	690	840	1,530
1505	8' deep	"	780	1,080	1,860
1510	10' deep	"	890	1,340	2,230
4200	Frames and covers, 24" diameter				
4210	300 lb	EA	51.00	410	461
4220	400 lb	"	57.00	430	487
4980	Steps for manholes				
5000	7" x 9"	EA	10.25	17.25	27.50
5020	8" x 9"	"	11.25	22.00	33.25
02530.40	**SANITARY SEWERS**				
0980	Clay				
1000	6" pipe	LF	9.50	9.41	18.91
2980	PVC				
3000	4" pipe	LF	7.12	4.05	11.17
3010	6" pipe	"	7.50	8.11	15.61
02540.10	**DRAINAGE FIELDS**				
0080	Perforated PVC pipe, for drain field				
0100	4" pipe	LF	6.33	2.71	9.04
0120	6" pipe	"	6.78	5.08	11.86

		UNIT	LABOR	MAT.	TOTAL
02540.50	**SEPTIC TANKS**				
0980	Septic tank, precast concrete				
1000	1000 gals	EA	480	1,020	1,500
1200	2000 gals	"	710	2,740	3,450
1310	Leaching pit, precast concrete, 72" diameter				
1320	3' deep	EA	360	780	1,140
1340	6' deep	"	410	1,370	1,780
1360	8' deep	"	480	1,740	2,220
02630.70	**UNDERDRAIN**				
1480	Drain tile, clay				
1500	6" pipe	LF	6.33	4.52	10.85
1520	8" pipe	"	6.62	7.21	13.83
1580	Porous concrete, standard strength				
1600	6" pipe	LF	6.33	5.14	11.47
1620	8" pipe	"	6.62	5.56	12.18
1800	Corrugated metal pipe, perforated type				
1810	6" pipe	LF	7.12	7.13	14.25
1820	8" pipe	"	7.50	8.43	15.93
1980	Perforated clay pipe				
2000	6" pipe	LF	8.14	5.98	14.12
2020	8" pipe	"	8.38	8.02	16.40
2480	Drain tile, concrete				
2500	6" pipe	LF	6.33	4.08	10.41
2520	8" pipe	"	6.62	6.35	12.97
4980	Perforated rigid PVC underdrain pipe				
5000	4" pipe	LF	4.75	2.10	6.85
5100	6" pipe	"	5.70	4.04	9.74
5150	8" pipe	"	6.33	6.17	12.50
6980	Underslab drainage, crushed stone				
7000	3" thick	SF	0.95	0.33	1.28
7120	4" thick	"	1.09	0.45	1.54
7140	6" thick	"	1.18	0.68	1.86
7180	Plastic filter fabric for drain lines	"	0.51	0.50	1.01
02740.20	**ASPHALT SURFACES**				
0050	Asphalt wearing surface, flexible pavement				
0100	1" thick	SY	2.83	4.52	7.35
0120	1-1/2" thick	"	3.40	6.82	10.22
1000	Binder course				
1010	1-1/2" thick	SY	3.15	6.45	9.60
1030	2" thick	"	3.87	8.58	12.45
2000	Bituminous sidewalk, no base				
2020	2" thick	SY	3.35	9.84	13.19
2040	3" thick	"	3.56	14.75	18.31
02750.10	**CONCRETE PAVING**				
1080	Concrete paving, reinforced, 5000 psi concrete				
2000	6" thick	SY	26.50	28.75	55.25
2005	7" thick	"	28.50	33.50	62.00
2010	8" thick	"	30.50	38.25	68.75
02810.40	**LAWN IRRIGATION**				
0480	Residential system, complete				
0500	Minimum	ACRE			19,250
0520	Maximum	"			36,630

DIVISION # 02 SITE CONSTRUCTION

		UNIT	LABOR	MAT.	TOTAL
02820.10	**CHAIN LINK FENCE**				
0230	Chain link fence, 9 ga., galvanized, with posts 10' o.c.				
0250	4' high	LF	3.64	7.38	11.02
0260	5' high	"	4.64	9.87	14.51
0270	6' high	"	6.38	11.25	17.63
1161	Fabric, galvanized chain link, 2" mesh, 9 ga.				
1163	4' high	LF	1.70	4.01	5.71
1164	5' high	"	2.04	4.91	6.95
1165	6' high	"	2.55	6.87	9.42
02820.20	**WOOD FENCE**				
0010	4x4 posts w/2x4 horizontals, 4'-8' high				
0020	Cedar, 1x4 planks, picket	SF	1.27	1.93	3.20
0030	1x6 planks, privacy	"	1.02	3.06	4.08
0040	1x8 planks, privacy	"	0.92	2.80	3.72
0050	Redwood, 1x4 planks, picket	"	1.27	1.57	2.84
0060	1x6 planks, privacy	"	1.02	2.91	3.93
0070	1x8 planks	"	0.92	2.54	3.46
0080	Treated pine, 1x4 planks, picket	"	1.27	1.55	2.82
0100	1x6 planks, privacy	"	1.02	1.51	2.53
0110	1x8 planks, privacy	"	0.92	1.48	2.40
0210	4' high				
0220	Composite, 1x4 planks, picket	SF	1.59	1.69	3.28
0230	1x6 planks, privacy	"	1.27	2.91	4.18
0240	1x8 planks	"	1.06	2.54	3.60
0250	Vinyl, 1x4 planks, picket	"	1.59	1.51	3.10
0260	1x6 planks, privacy	"	1.27	2.68	3.95
0270	1x8 planks, privacy	"	1.06	2.64	3.70
0280	Gate, cedar or redwood, 3' w, 1x4 planks, picket	EA	42.50	26.50	69.00
0290	1x6 planks, privacy	"	51.00	73.00	124
0300	1x8 planks, privacy	"	57.00	83.00	140
02840.40	**PARKING BARRIERS**				
3010	Bollard, concrete filled, 8' long				
3020	6" dia.	EA	43.50	600	644
3030	8" dia.	"	54.00	890	944
3040	12" dia.	"	65.00	1,140	1,205
02880.70	**RECREATIONAL COURTS**				
1000	Walls, galvanized steel				
1020	8' high	LF	10.25	16.00	26.25
1040	10' high	"	11.25	18.75	30.00
1060	12' high	"	13.50	21.75	35.25
1200	Vinyl coated				
1220	8' high	LF	10.25	15.25	25.50
1240	10' high	"	11.25	18.75	30.00
1260	12' high	"	13.50	20.75	34.25
2010	Gates, galvanized steel				
2200	Single, 3' transom				
2210	3'x7'	EA	260	370	630
2220	4'x7'	"	290	390	680
2230	5'x7'	"	340	540	880
2240	6'x7'	"	410	580	990
2400	Vinyl coated				
2405	Single, 3' transom				

		UNIT	LABOR	MAT.	TOTAL
02880.70	**RECREATIONAL COURTS, Cont'd...**				
2410	3'x7'	EA	260	730	990
2420	4'x7'	"	290	790	1,080
2430	5'x7'	"	340	790	1,130
2440	6'x7'	"	410	820	1,230
02910.10	**TOPSOIL**				
0005	Spread topsoil, with equipment				
0010	Minimum	CY	15.50		15.50
0020	Maximum	"	19.50		19.50
0080	By hand				
0100	Minimum	CY	51.00		51.00
0110	Maximum	"	64.00		64.00
0980	Area prep. seeding (grade, rake and clean)				
1000	Square yard	SY	0.40		0.40
1020	By acre	ACRE	2,040		2,040
2000	Remove topsoil and stockpile on site				
2020	4" deep	CY	13.00		13.00
2040	6" deep	"	12.00		12.00
2200	Spreading topsoil from stock pile				
2220	By loader	CY	14.25		14.25
2240	By hand	"	160		160
2260	Top dress by hand	SY	1.55		1.55
2280	Place imported top soil				
2300	By loader				
2320	4" deep	SY	1.55		1.55
2340	6" deep	"	1.72		1.72
2360	By hand				
2370	4" deep	SY	5.67		5.67
2380	6" deep	"	6.38		6.38
5980	Plant bed preparation, 18" deep				
6000	With backhoe/loader	SY	3.88		3.88
6010	By hand	"	8.51		8.51
02920.10	**FERTILIZING**				
0080	Fertilizing (23#/1000 sf)				
0100	By square yard	SY	0.17	0.03	0.20
0120	By acre	ACRE	850	190	1,040
2980	Liming (70#/1000 sf)				
3000	By square yard	SY	0.22	0.03	0.25
3020	By acre	ACRE	1,130	190	1,320
02920.30	**SEEDING**				
0980	Mechanical seeding, 175 lb/acre				
1000	By square yard	SY	0.13	0.23	0.36
1020	By acre	ACRE	680	930	1,610
2040	450 lb/acre				
2060	By square yard	SY	0.17	0.59	0.76
2080	By acre	ACRE	850	2,310	3,160
5980	Seeding by hand, 10 lb per 100 SY				
6000	By square yard	SY	0.17	0.66	0.83
6010	By acre	ACRE	850	2,580	3,430
8010	Reseed disturbed areas	SF	0.25	0.06	0.31

		UNIT	LABOR	MAT.	TOTAL
02935.10	**SHRUB & TREE MAINTENANCE**				
1000	Moving shrubs on site				
1220	3' high	EA	51.00		51.00
1240	4' high	"	57.00		57.00
2000	Moving trees on site				
3060	6' high	EA	63.00		63.00
3080	8' high	"	71.00		71.00
3100	10' high	"	95.00		95.00
3110	Palm trees				
3140	10' high	EA	95.00		95.00
3144	40' high	"	570		570
02935.30	**WEED CONTROL**				
1000	Weed control, bromicil, 15 lb./acre, wettable powder	ACRE	260	310	570
1100	Vegetation control, by application of plant killer	SY	0.20	0.02	0.22
1200	Weed killer, lawns and fields	"	0.10	0.26	0.36
02945.20	**LANDSCAPE ACCESSORIES**				
0100	Steel edging, 3/16" x 4"	LF	0.63	1.29	1.92
0200	Landscaping stepping stones, 15"x15", white	EA	2.55	5.83	8.38
6000	Wood chip mulch	CY	34.00	40.75	74.75
6010	2" thick	SY	1.02	2.49	3.51
6020	4" thick	"	1.45	4.70	6.15
6030	6" thick	"	1.85	7.04	8.89
6200	Gravel mulch, 3/4" stone	CY	51.00	32.25	83.25
6300	White marble chips, 1" deep	SF	0.51	0.63	1.14
6980	Peat moss				
7000	2" thick	SY	1.13	3.46	4.59
7020	4" thick	"	1.70	6.66	8.36
7030	6" thick	"	2.12	10.25	12.37
7980	Landscaping timbers, treated lumber				
8000	4" x 4"	LF	1.70	3.58	5.28
8020	6" x 6"	"	1.82	8.32	10.14
8040	8" x 8"	"	2.12	10.00	12.12

		UNIT	COST
02999.10	**DEMOLITION**		
3200	Selective Building Removals		
3210	No Cutting or Disposal Included		
3220	Concrete - Hand Work		
3230	8" Walls - Reinforced	SF	17.75
3240	Non - Reinforced	"	11.75
3250	12" Walls - Reinforced	"	28.50
3260	12" Footings x 24" wide	LF	26.00
3270	x 36" wide	"	35.75
3280	16" Footings x 24" wide	SF	47.50
3290	6" Structural Slab - Reinforced	"	11.00
3300	8" Structural Slab - Reinforced	"	13.50
3310	4" Slab on Ground - Reinforced	"	5.09
3320	Non - Reinforced	"	3.80
3330	6" Slab on Ground - Reinforced	"	6.06
3340	Non - Reinforced	"	4.52
3350	Stairs - Reinforced	"	19.00
3360	Masonry - Hand Work		
3370	4" Brick or Stone Walls	SF	3.43
3380	4" Brick and 8" Backup Block or Tile	"	6.33
3390	4" Block or Tile Partitions	"	3.35
3400	6" Block or Tile Partitions	"	3.65
3410	8" Block or Tile Partitions	"	4.19
3420	12" Block or Tile Partitions	"	5.27
3430	Miscellaneous - Hand Work		
3440	Acoustical Ceilings - Attached	SF	0.98
3450	Suspended (Including Grid)	"	0.56
3460	Asbestos - Pipe	LF	71.00
3470	Ceilings and Walls	SF	23.75
3480	Columns and Beams	"	57.00
3490	Tile Flooring	"	2.66
3500	Cabinets and Tops	LF	15.75
3510	Carpet	SF	0.49
3520	Ceramic and Quarry Tile	"	1.90
3530	Doors and Frames - Metal	EA	84.00
3540	Wood	"	71.00
3550	Drywall Ceilings - Attached	SF	1.19
3560	Drywall on Wood or Metal Studs - 2 Sides	"	1.33
3570	Paint Removal - Doors and Windows	"	1.19
3580	Walls	"	0.98
3590	Plaster Ceilings - Attached (Including Iron)	"	1.68
3600	on Wood or Metal Studs	"	1.61
3610	Roofing - Builtup	"	1.61
3620	Shingles - Asphalt and Wood	"	0.56
3630	Terrazzo Flooring	"	2.66
3640	Vinyl Flooring	"	0.56
3650	Wall Coverings	"	1.05
3660	Windows	EA	57.00
3670	Wood Flooring	SF	0.56
4300	Site Removals (Including Loading)		
4310	4" Concrete Walks - Labor Only (Non-Reinforced)	SF	3.80
4320	Machine Only (Non-Reinforced)	"	1.26
4330	6" Concrete Drives - Labor Only (Non-Reinforced)	"	4.75

		UNIT	COST
02999.10	**DEMOLITION, Cont'd...**		
4340	Machine Only (Reinforced)	SF	1.68
4350	6" x 18" Concrete Curb - Machine	LF	3.23
4360	Curb and Gutter - Machine	"	4.21
4370	2" Asphalt - Machine	SF	0.73
4380	Fencing - 8' Hand	LF	4.35
02999.20	**EARTHWORK**		
1000	GRADING - Hand - 4" - Site	SF	0.43
1010	Hand – 4" Building	"	0.73
2000	EXCAVATION - Hand - Open - Soft (Sand)	CY	57.00
2010	Medium (Clay)	"	71.00
2020	Hard (Shale)	"	110
2030	Add for Trench or Pocket	PCT	15.00
3000	BACK FILL - Hand - Not Comp. (Site Borrow)	CY	26.00
4000	BACK FILL - Hand - Comp.(Site Borrow)		
4010	12" Lifts - Building - No Machine	CY	43.75
4020	With Machine	"	35.75
4030	18" Lifts - Building - No Machine	"	40.75
4040	With Machine	"	33.50
02999.50	**DRAINAGE**		
4000	BUILDING FOUNDATION DRAINAGE		
4010	4" Clay Pipe	LF	10.42
4020	4" Plastic Pipe - Perforated	"	8.38
4030	6" Clay Pipe	"	11.77
4040	6" Plastic Pipe - Perforated	"	13.26
4050	Add for Porous Surround - 2' x 2'	"	15.33
02999.60	**PAVEMENT, CURBS AND WALKS**		
2000	CURBS AND GUTTERS		
2100	Concrete - Cast-in-Place (Machine Placed)		
2110	Curb - 6" x 12"	LF	24.97
2120	6" x 18"	"	29.29
2130	6" x 24"	"	33.49
2140	6" x 30"	"	38.31
2150	Curb and Gutter - 6" x 12"	"	36.18
2160	6" x 18"	"	42.75
2170	6" x 24"	"	49.00
2180	Add for Hand Placed	"	11.75
2190	Add for 2 #5 Reinf. Rods	"	5.16
2191	Add for Curves and Radius Work	PCT	44.00
2200	Concrete Precast - 6" x 10" x 8"	LF	26.97
2210	6" x 9" x 8"	"	24.60
2250	6" x 8"	"	21.78
2300	Bituminous - 6" x 8"	"	9.59
2400	Granite - 6" x 16"	"	79.00
2500	Timbers - Treated - 6" x 6"	"	17.84
2600	Plastic - 6" x 6"	"	23.47
3000	WALKS		
3010	Bituminous - 1-1/2" with 4" Sand Base	SF	4.70
3020	2" with 4" Sand Base	"	5.17
3030	Concrete - 4" - Broom Finish	"	8.67
3040	5" - Broom Finish	"	9.80
3050	6" - Broom Finish	"	11.05

		UNIT	COST
02999.60	**PAVEMENT, CURBS AND WALKS, Cont'd...**		
3060	Add for 6" x 6", 10 - 10 Mesh	SF	1.93
3070	Add for 4" Sand Base	"	1.70
3080	Add for Exposed Aggregate	"	2.70
3090	Crushed Rock - 4"	"	1.85
4000	Brick - 4" - with 2" Sand Cushion	"	21.27
4010	4" - with 2" Mortar Setting Bed	"	25.29
4020	Flagstone – 1-1/4" - with 4" Sand Cushion	"	34.64
4030	1-1/4" - with 2" Mortar Setting Bed	"	38.75
4040	Precast Block - 1" Colored with 4" Sand Cushion	"	9.46
4050	2" Colored with 4" Sand Cushion	"	10.97
4060	Wood - 2" Boards on 6" x 6" Timbers	"	16.69
4070	2" Boards on 4" x 4" Timbers	"	14.13
4080	Slate - 1-1/4" - with 2" Mortar Setting Bed	"	46.25

TABLE OF CONTENTS   PAGE

		UNIT	LABOR	MAT.	TOTAL
03100.03	**FORMWORK ACCESSORIES**				
1000	Column clamps				
1010	Small, adjustable, 24"x24"	EA			87.00
1020	Medium 36"x36"	"			90.00
1030	Large 60"x60"	"			91.00
2000	Forming hangers				
2010	Iron 14 ga.	EA			2.81
2020	22 ga.	"			2.81
3000	Snap ties				
3010	Short-end with washers, 6" long	EA			1.73
3020	12" long	"			1.96
4000	18" long	"			2.34
4010	24" long	"			2.52
4020	Long-end with washers, 6' long	"			2.04
4030	12" long	"			2.26
4040	18" long	"			2.56
4050	24" long	"			2.87
5000	Stakes				
5010	Round, pre-drilled holes, 12" long	EA			6.14
5020	18" long	"			6.87
5030	24" long	"			8.90
5040	30" long	"			11.50
5050	36" long	"			13.75
5060	48" long	"			18.50
5070	I beam type, 12" long	"			5.05
6000	18" long	"			5.75
6010	24" long	"			8.73
6020	30" long	"			10.25
6030	36" long	"			13.00
7000	48" long	"			16.25
7010	Taper ties				
7020	50K, 1-1/4" to 1", 35" long	EA			85.00
8000	45" long	"			140
8010	55" long	"			170
8020	Walers				
8030	5" deep, 4' long	EA			240
8040	8' long	"			310
8050	12' long	"			510
9000	16' long	"			660
9010	8" deep, 4' long	"			320
9020	8' long	"			580
9030	12' long	"			840
9040	16' long	"			1,330
03110.05	**BEAM FORMWORK**				
1000	Beam forms, job built				
1020	Beam bottoms				
1040	1 use	SF	10.75	5.15	15.90
1080	3 uses	"	10.00	2.32	12.32
1120	5 uses	"	9.32	1.75	11.07
2000	Beam sides				
2020	1 use	SF	7.24	3.68	10.92
2060	3 uses	"	6.52	1.92	8.44

		UNIT	LABOR	MAT.	TOTAL
03110.05	**BEAM FORMWORK, Cont'd...**				
2100	5 uses	SF	5.93	1.56	7.49
03110.15	**COLUMN FORMWORK**				
1000	Column, square forms, job built				
1020	8" x 8" columns				
1040	1 use	SF	13.00	4.33	17.33
1080	3 uses	"	12.00	1.97	13.97
1120	5 uses	"	11.25	1.53	12.78
1300	16" x 16" columns				
1320	1 use	SF	10.75	3.77	14.52
1360	3 uses	"	10.25	1.59	11.84
1390	5 uses	"	9.59	1.19	10.78
2000	Round fiber forms, 1 use				
2040	10" dia.	LF	13.00	5.65	18.65
2120	18" dia.	"	15.50	19.50	35.00
2180	36" dia.	"	19.75	44.50	64.25
03110.18	**CURB FORMWORK**				
0980	Curb forms				
0990	Straight, 6" high				
1000	1 use	LF	6.52	2.58	9.10
1040	3 uses	"	5.93	1.16	7.09
1080	5 uses	"	5.43	0.94	6.37
1090	Curved, 6" high				
2000	1 use	LF	8.15	2.79	10.94
2040	3 uses	"	7.24	1.34	8.58
2080	5 uses	"	6.65	1.13	7.78
03110.20	**ELEVATED SLAB FORMWORK**				
0100	Elevated slab formwork				
1000	Slab, with drop panels				
1020	1 use	SF	5.21	4.63	9.84
1060	3 uses	"	4.83	2.08	6.91
1100	5 uses	"	4.49	1.66	6.15
2000	Floor slab, hung from steel beams				
2020	1 use	SF	5.01	3.73	8.74
2060	3 uses	"	4.66	1.86	6.52
2100	5 uses	"	4.34	1.37	5.71
3000	Floor slab, with pans or domes				
3020	1 use	SF	5.93	6.65	12.58
3060	3 uses	"	5.43	4.03	9.46
3100	5 uses	"	5.01	3.33	8.34
9030	Equipment curbs, 12" high				
9035	1 use	LF	6.52	3.43	9.95
9060	3 uses	"	5.93	1.92	7.85
9100	5 uses	"	5.43	1.49	6.92
03110.25	**EQUIPMENT PAD FORMWORK**				
1000	Equipment pad, job built				
1020	1 use	SF	8.15	4.50	12.65
1040	2 uses	"	7.67	2.70	10.37
1060	3 uses	"	7.24	2.16	9.40

		UNIT	LABOR	MAT.	TOTAL
03110.35	**FOOTING FORMWORK**				
2000	Wall footings, job built, continuous				
2040	1 use	SF	6.52	2.09	8.61
2060	3 uses	"	5.93	1.21	7.14
2090	5 uses	"	5.43	0.93	6.36
3000	Column footings, spread				
3020	1 use	SF	8.15	2.21	10.36
3060	3 uses	"	7.24	1.17	8.41
3100	5 uses	"	6.52	0.90	7.42
03110.50	**GRADE BEAM FORMWORK**				
1000	Grade beams, job built				
1020	1 use	SF	6.52	3.29	9.81
1060	3 uses	"	5.93	1.44	7.37
1100	5 uses	"	5.43	1.00	6.43
03110.53	**PILE CAP FORMWORK**				
1500	Pile cap forms, job built				
1510	Square				
1520	1 use	SF	8.15	3.74	11.89
1560	3 uses	"	7.24	1.71	8.95
1600	5 uses	"	6.52	1.25	7.77
03110.55	**SLAB / MAT FORMWORK**				
3000	Mat foundations, job built				
3020	1 use	SF	8.15	3.27	11.42
3060	3 uses	"	7.24	1.39	8.63
3100	5 uses	"	6.52	0.94	7.46
3980	Edge forms				
3990	6" high				
4000	1 use	LF	5.93	3.30	9.23
4002	3 uses	"	5.43	1.39	6.82
4004	5 uses	"	5.01	0.95	5.96
4014	5 uses	"	5.43	0.86	6.29
5000	Formwork for openings				
5020	1 use	SF	13.00	4.45	17.45
5060	3 uses	"	10.75	2.14	12.89
5100	5 uses	"	9.32	1.38	10.70
03110.60	**STAIR FORMWORK**				
1000	Stairway forms, job built				
1020	1 use	SF	13.00	5.14	18.14
1040	3 uses	"	10.75	2.24	12.99
1060	5 uses	"	9.32	1.72	11.04
03110.65	**WALL FORMWORK**				
2980	Wall forms, exterior, job built				
3000	Up to 8' high wall				
3120	1 use	SF	6.52	3.52	10.04
3160	3 uses	"	5.93	1.71	7.64
3190	5 uses	"	5.43	1.29	6.72
3200	Over 8' high wall				
3220	1 use	SF	8.15	3.87	12.02
3240	3 uses	"	7.24	2.01	9.25
3290	5 uses	"	6.52	1.58	8.10
5000	Column pier and pilaster				
5020	1 use	SF	13.00	3.87	16.87

		UNIT	LABOR	MAT.	TOTAL
03110.65	**WALL FORMWORK, Cont'd...**				
5060	3 uses	SF	10.75	2.13	12.88
5090	5 uses	"	9.32	1.74	11.06
6980	Interior wall forms				
7000	Up to 8' high				
7020	1 use	SF	5.93	3.52	9.45
7060	3 uses	"	5.43	1.74	7.17
7100	5 uses	"	5.01	1.25	6.26
7200	Over 8' high				
7220	1 use	SF	7.24	3.87	11.11
7260	3 uses	"	6.52	2.01	8.53
7290	5 uses	"	5.93	1.60	7.53
9000	PVC form liner, per side, smooth finish				
9010	1 use	SF	5.43	9.09	14.52
9030	3 uses	"	5.01	4.22	9.23
9050	5 uses	"	4.34	2.60	6.94
03110.70	**INSULATED CONCRETE FORMS**				
0010	4" thick, straight	SF	2.03	5.24	7.27
0020	90° corner	"	2.03	5.24	7.27
0030	45° angle	"	2.03	5.75	7.78
1000	6" thick, straight	"	2.03	5.31	7.34
1010	90° corner	"	2.03	5.31	7.34
1020	45° angle	"	2.03	5.75	7.78
1030	6" thick, corbel ledge	"	2.03	6.62	8.65
1040	T-block	"	2.03	5.94	7.97
2000	8" thick, straight	"	2.03	5.47	7.50
2010	90° corner	"	2.03	5.31	7.34
2020	45° angle	"	2.03	6.11	8.14
2030	corbel ledge	"	2.03	6.83	8.86
2040	T-block	"	2.03	6.28	8.31
03110.90	**MISCELLANEOUS FORMWORK**				
1200	Keyway forms (5 uses)				
1220	2 x 4	LF	3.26	0.30	3.56
1240	2 x 6	"	3.62	0.43	4.05
1500	Bulkheads				
1510	Walls, with keyways				
1515	2 piece	LF	5.93	4.97	10.90
1520	3 piece	"	6.52	6.28	12.80
1560	Elevated slab, with keyway				
1570	2 piece	LF	5.43	5.69	11.12
1580	3 piece	"	5.93	8.39	14.32
1600	Ground slab, with keyway				
1620	2 piece	LF	4.66	5.88	10.54
1640	3 piece	"	5.01	7.19	12.20
2000	Chamfer strips				
2020	Wood				
2040	1/2" wide	LF	1.44	0.28	1.72
2060	3/4" wide	"	1.44	0.36	1.80
2070	1" wide	"	1.44	0.49	1.93
2100	PVC				
2120	1/2" wide	LF	1.44	1.26	2.70
2140	3/4" wide	"	1.44	1.36	2.80

		UNIT	LABOR	MAT.	TOTAL
03110.90	**MISCELLANEOUS FORMWORK, Cont'd...**				
2160	1" wide	LF	1.44	1.98	3.42
2170	Radius				
2180	1"	LF	1.55	1.47	3.02
2200	1-1/2"	"	1.55	2.67	4.22
3000	Reglets				
3020	Galvanized steel, 24 ga.	LF	2.60	1.91	4.51
5000	Metal formwork				
5020	Straight edge forms				
5080	8" high	LF	4.66	35.25	39.91
5300	Curb form, S-shape				
5310	12" x				
5340	2'	LF	8.69	57.00	65.69
5380	3'	"	7.24	67.00	74.24
03210.05	**BEAM REINFORCING**				
0980	Beam-girders				
1000	#3 - #4	TON	1,660	1,980	3,640
1011	#7 - #8	"	1,110	1,650	2,760
1018	Galvanized				
1020	#3 - #4	TON	1,660	3,360	5,020
1031	#7 - #8	"	1,110	3,060	4,170
2000	Bond Beams				
2100	#3 - #4	TON	2,220	1,980	4,200
2120	#7 - #8	"	1,480	1,650	3,130
2200	Galvanized				
2210	#3 - #4	TON	2,220	3,220	5,440
2230	#7 - #8	"	1,480	3,060	4,540
03210.15	**COLUMN REINFORCING**				
0980	Columns				
1000	#3 - #4	TON	1,900	1,980	3,880
1015	#7 - #8	"	1,330	1,650	2,980
1100	Galvanized				
1200	#3 - #4	TON	1,900	3,360	5,260
1320	#7 - #8	"	1,330	3,060	4,390
03210.20	**ELEVATED SLAB REINFORCING**				
0980	Elevated slab				
1000	#3 - #4	TON	830	1,980	2,810
1040	#7 - #8	"	670	1,650	2,320
1980	Galvanized				
2000	#3 - #4	TON	830	3,220	4,050
2040	#7 - #8	"	670	3,060	3,730
03210.25	**EQUIP. PAD REINFORCING**				
0980	Equipment pad				
1000	#3 - #4	TON	1,330	1,980	3,310
1040	#7 - #8	"	1,110	1,650	2,760
03210.35	**FOOTING REINFORCING**				
1000	Footings				
1010	Grade 50				
1020	#3 - #4	TON	1,110	1,980	3,090
1040	#7 - #8	"	830	1,650	2,480
1055	Grade 60				
1060	#3 - #4	TON	1,110	1,980	3,090

		UNIT	LABOR	MAT.	TOTAL
03210.35	**FOOTING REINFORCING, Cont'd...**				
1074	#7 - #8	TON	830	1,650	2,480
4980	Straight dowels, 24" long				
5000	1" dia. (#8)	EA	1.66	6.13	7.79
5040	3/4" dia. (#6)	"	1.54	5.51	7.05
5050	5/8" dia. (#5)	"	1.44	4.76	6.20
5060	1/2" dia. (#4)	"	1.33	3.59	4.92
03210.45	**FOUNDATION REINFORCING**				
0980	Foundations				
1000	#3 - #4	TON	1,110	1,980	3,090
1040	#7 - #8	"	830	1,650	2,480
1380	Galvanized				
1400	#3 - #4	TON	1,110	3,370	4,480
1420	#7 - #8	"	830	3,070	3,900
03210.50	**GRADE BEAM REINFORCING**				
0980	Grade beams				
1000	#3 - #4	TON	1,020	1,980	3,000
1040	#7 - #8	"	780	1,650	2,430
1090	Galvanized				
1100	#3 - #4	TON	1,020	3,370	4,390
1140	#7 - #8	"	780	3,070	3,850
03210.53	**PILE CAP REINFORCING**				
0980	Pile caps				
1000	#3 - #4	TON	1,660	1,980	3,640
1040	#7 - #8	"	1,330	1,650	2,980
1090	Galvanized				
1100	#3 - #4	TON	1,660	3,370	5,030
1140	#7 - #8	"	1,330	3,070	4,400
03210.55	**SLAB / MAT REINFORCING**				
0980	Bars, slabs				
1000	#3 - #4	TON	1,110	1,980	3,090
1040	#7 - #8	"	830	1,650	2,480
1980	Galvanized				
2000	#3 - #4	TON	1,110	3,370	4,480
2020	#5 - #6	"	950	3,190	4,140
2040	#7 - #8	"	830	3,070	3,900
5000	Wire mesh, slabs				
5010	Galvanized				
5015	4x4				
5020	W1.4xW1.4	SF	0.44	0.49	0.93
5040	W2.0xW2.0	"	0.47	0.63	1.10
5060	W2.9xW2.9	"	0.51	0.89	1.40
5080	W4.0xW4.0	"	0.55	1.32	1.87
5090	6x6				
5100	W1.4xW1.4	SF	0.33	0.45	0.78
5120	W2.0xW2.0	"	0.37	0.63	1.00
5140	W2.9xW2.9	"	0.39	0.86	1.25
5160	W4.0xW4.0	"	0.44	0.93	1.37
5170	Standard				
5175	2x2				
5180	W.9xW.9	SF	0.44	0.49	0.93
5185	4x4				

		UNIT	LABOR	MAT.	TOTAL
03210.55	**SLAB / MAT REINFORCING, Cont'd...**				
5190	W1.4xW1.4	SF	0.44	0.32	0.76
5500	W4.0xW4.0	"	0.55	0.89	1.44
5580	6x6				
5600	W1.4xW1.4	SF	0.33	0.21	0.54
6020	W4.0xW4.0	"	0.44	0.60	1.04
03210.60	**STAIR REINFORCING**				
0980	Stairs				
1000	#3 - #4	TON	1,330	1,980	3,310
1020	#5 - #6	"	1,110	1,740	2,850
1980	Galvanized				
2000	#3 - #4	TON	1,330	3,370	4,700
2020	#5 - #6	"	1,110	3,190	4,300
03210.65	**WALL REINFORCING**				
0980	Walls				
1000	#3 - #4	TON	950	1,980	2,930
1040	#7 - #8	"	740	1,650	2,390
1980	Galvanized				
2000	#3 - #4	TON	950	3,370	4,320
2040	#7 - #8	"	740	3,070	3,810
8980	Masonry wall (horizontal)				
9000	#3 - #4	TON	2,660	1,980	4,640
9020	#5 - #6	"	2,220	1,740	3,960
9030	Galvanized				
9040	#3 - #4	TON	2,660	3,370	6,030
9060	#5 - #6	"	2,220	3,190	5,410
9180	Masonry wall (vertical)				
9200	#3 - #4	TON	3,330	1,980	5,310
9220	#5 - #6	"	2,660	1,740	4,400
9230	Galvanized				
9240	#3 - #4	TON	3,330	3,370	6,700
9260	#5 - #6	"	2,660	3,190	5,850
03250.40	**CONCRETE ACCESSORIES**				
1000	Expansion joint, poured				
1010	Asphalt				
1020	1/2" x 1"	LF	1.02	0.88	1.90
1040	1" x 2"	"	1.11	2.76	3.87
1060	Liquid neoprene, cold applied				
1080	1/2" x 1"	LF	1.04	3.45	4.49
1100	1" x 2"	"	1.13	14.25	15.38
03300.10	**CONCRETE ADMIXTURES**				
1000	Concrete admixtures				
1020	Water reducing admixture	GAL			12.00
1040	Set retarder	"			25.75
1060	Air entraining agent	"			11.25
03350.10	**CONCRETE FINISHES**				
0980	Floor finishes				
1000	Broom	SF	0.72		0.72
1020	Screed	"	0.63		0.63
1040	Darby	"	0.63		0.63
1060	Steel float	"	0.85		0.85
4000	Wall finishes				

DIVISION # 03 CONCRETE

		UNIT	LABOR	MAT.	TOTAL
03350.10	**CONCRETE FINISHES, Cont'd...**				
4020	Burlap rub, with cement paste	SF	0.85	0.12	0.97
4160	Break ties and patch holes	"	1.02		1.02
4170	Carborundum				
4180	Dry rub	SF	1.70		1.70
4200	Wet rub	"	2.55		2.55
5000	Floor hardeners				
5010	Metallic				
5020	Light service	SF	0.63	0.43	1.06
5040	Heavy service	"	0.85	1.30	2.15
5050	Non-metallic				
5060	Light service	SF	0.63	0.21	0.84
5080	Heavy service	"	0.85	0.90	1.75
03360.10	**PNEUMATIC CONCRETE**				
0100	Pneumatic applied concrete (gunite)				
1035	2" thick	SF	3.56	6.33	9.89
1040	3" thick	"	4.75	7.77	12.52
1060	4" thick	"	5.70	9.48	15.18
1980	Finish surface				
2000	Minimum	SF	3.26		3.26
2020	Maximum	"	6.52		6.52
03370.10	**CURING CONCRETE**				
1000	Sprayed membrane				
1010	Slabs	SF	0.10	0.06	0.16
1020	Walls	"	0.12	0.08	0.20
1025	Curing paper				
1030	Slabs	SF	0.13	0.08	0.21
2000	Walls	"	0.15	0.08	0.23
2010	Burlap				
2020	7.5 oz.	SF	0.17	0.07	0.24
2500	12 oz.	"	0.18	0.10	0.28
03380.05	**BEAM CONCRETE**				
0960	Beams and girders				
0980	2500# or 3000# concrete				
1010	By pump	CY	94.00	130	224
1020	By hand buggy	"	51.00	130	181
4000	5000# concrete				
4020	By pump	CY	94.00	150	244
4040	By hand buggy	"	51.00	150	201
9460	Bond beam, 3000# concrete				
9470	By pump				
9480	8" high				
9500	4" wide	LF	2.06	0.36	2.42
9520	6" wide	"	2.34	0.89	3.23
9530	8" wide	"	2.58	1.13	3.71
9540	10" wide	"	2.86	1.51	4.37
9550	12" wide	"	3.22	2.03	5.25
03380.15	**COLUMN CONCRETE**				
0980	Columns				
0990	2500# or 3000# concrete				
1010	By pump	CY	86.00	140	226
3980	5000# concrete				

		UNIT	LABOR	MAT.	TOTAL
03380.15	**COLUMN CONCRETE, Cont'd...**				
4020	By pump	CY	86.00	150	236
03380.25	**EQUIPMENT PAD CONCRETE**				
0960	Equipment pad				
0980	2500# or 3000# concrete				
1000	By chute	CY	17.00	140	157
1020	By pump	"	74.00	140	214
1050	3500# or 4000# concrete				
1060	By chute	CY	17.00	140	157
1080	By pump	"	74.00	140	214
1110	5000# concrete				
1120	By chute	CY	17.00	150	167
1140	By pump	"	74.00	150	224
03380.35	**FOOTING CONCRETE**				
0980	Continuous footing				
0990	2500# or 3000# concrete				
1000	By chute	CY	17.00	140	157
1010	By pump	"	65.00	140	205
4000	5000# concrete				
4010	By chute	CY	17.00	150	167
4020	By pump	"	65.00	150	215
4980	Spread footing				
5000	2500# or 3000# concrete				
5010	Under 5 c.y.				
5020	By chute	CY	17.00	130	147
5040	By pump	"	69.00	130	199
7200	5000# concrete				
7205	Under 5 c.y.				
7210	By chute	CY	17.00	150	167
7220	By pump	"	69.00	150	219
03380.50	**GRADE BEAM CONCRETE**				
0960	Grade beam				
0980	2500# or 3000# concrete				
1000	By chute	CY	17.00	130	147
1040	By pump	"	65.00	130	195
1060	By hand buggy	"	51.00	130	181
1150	5000# concrete				
1160	By chute	CY	17.00	150	167
1190	By pump	"	65.00	150	215
1200	By hand buggy	"	51.00	150	201
03380.53	**PILE CAP CONCRETE**				
0970	Pile cap				
0980	2500# or 3000 concrete				
1000	By chute	CY	17.00	140	157
1010	By pump	"	74.00	140	214
1020	By hand buggy	"	51.00	140	191
3980	5000# concrete				
4010	By chute	CY	17.00	150	167
4020	By pump	"	74.00	150	224
4030	By hand buggy	"	51.00	150	201

		UNIT	LABOR	MAT.	TOTAL
03380.55	**SLAB / MAT CONCRETE**				
0960	Slab on grade				
0980	2500# or 3000# concrete				
1000	By chute	CY	12.75	140	153
1020	By pump	"	37.00	140	177
1030	By hand buggy	"	34.00	140	174
3980	5000# concrete				
4010	By chute	CY	12.75	150	163
4030	By pump	"	37.00	150	187
4040	By hand buggy	"	34.00	150	184
03380.58	**SIDEWALKS**				
6000	Walks, with wire mesh, base not incl.				
6010	4" thick	SF	1.70	1.93	3.63
6020	5" thick	"	2.04	2.61	4.65
6030	6" thick	"	2.55	3.21	5.76
03380.60	**STAIR CONCRETE**				
0960	Stairs				
0980	2500# or 3000# concrete				
1000	By chute	CY	17.00	140	157
1030	By pump	"	74.00	140	214
1040	By hand buggy	"	51.00	140	191
2100	3500# or 4000# concrete				
2120	By chute	CY	17.00	140	157
2160	By pump	"	74.00	140	214
2180	By hand buggy	"	51.00	140	191
4000	5000# concrete				
4010	By chute	CY	17.00	150	167
4030	By pump	"	74.00	150	224
4040	By hand buggy	"	51.00	150	201
03380.65	**WALL CONCRETE**				
0940	Walls				
0960	2500# or 3000# concrete				
0980	To 4'				
1000	By chute	CY	14.50	140	155
1010	By pump	"	79.00	140	219
1020	To 8'				
1040	By pump	CY	86.00	140	226
2960	3500# or 4000# concrete				
2980	To 4'				
3000	By chute	CY	14.50	140	155
3030	By pump	"	79.00	140	219
3060	To 8'				
3100	By pump	CY	86.00	140	226
8480	Filled block (CMU)				
8490	3000# concrete, by pump				
8500	4" wide	SF	3.68	0.51	4.19
8510	6" wide	"	4.30	1.15	5.45
8520	8" wide	"	5.16	1.80	6.96
8530	10" wide	"	6.07	2.41	8.48
8540	12" wide	"	7.37	3.10	10.47
8560	Pilasters, 3000# concrete	CF	100	7.04	107
8700	Wall cavity, 2" thick, 3000# concrete	SF	3.44	1.31	4.75

		UNIT	LABOR	MAT.	TOTAL
03550.10	**CONCRETE TOPPINGS**				
1000	Gypsum fill				
1020	2" thick	SF	0.53	2.11	2.64
1040	2-1/2" thick	"	0.54	2.40	2.94
1060	3" thick	"	0.56	2.96	3.52
1080	3-1/2" thick	"	0.57	3.38	3.95
1100	4" thick	"	0.64	3.95	4.59
2000	Formboard				
2020	Mineral fiber board				
2040	1" thick	SF	1.27	1.89	3.16
2060	1-1/2" thick	"	1.45	4.96	6.41
2070	Cement fiber board				
2080	1" thick	SF	1.70	1.47	3.17
2100	1-1/2" thick	"	1.96	1.89	3.85
2110	Glass fiber board				
2120	1" thick	SF	1.27	2.32	3.59
2140	1-1/2" thick	"	1.45	3.15	4.60
4000	Poured deck				
4010	Vermiculite or perlite				
4020	1 to 4 mix	CY	86.00	200	286
4040	1 to 6 mix	"	79.00	180	259
4050	Vermiculite or perlite				
4060	2" thick				
4080	1 to 4 mix	SF	0.54	1.89	2.43
4100	1 to 6 mix	"	0.49	1.38	1.87
4200	3" thick				
4220	1 to 4 mix	SF	0.79	2.58	3.37
4240	1 to 6 mix	"	0.73	2.04	2.77
6000	Concrete plank, lightweight				
6020	2" thick	SF	4.25	10.25	14.50
6040	2-1/2" thick	"	4.25	10.50	14.75
6080	3-1/2" thick	"	4.73	11.00	15.73
6100	4" thick	"	4.73	11.25	15.98
6500	Channel slab, lightweight, straight				
6520	2-3/4" thick	SF	4.25	8.28	12.53
6540	3-1/2" thick	"	4.25	8.52	12.77
6560	3-3/4" thick	"	4.25	9.20	13.45
6580	4-3/4" thick	"	4.73	11.75	16.48
7000	Gypsum plank				
7020	2" thick	SF	4.25	3.83	8.08
7040	3" thick	"	4.25	4.02	8.27
8000	Cement fiber, T & G planks				
8020	1" thick	SF	3.87	2.11	5.98
8040	1-1/2" thick	"	3.87	2.24	6.11
8060	2" thick	"	4.25	2.68	6.93
8080	2-1/2" thick	"	4.25	2.84	7.09
8100	3" thick	"	4.25	3.69	7.94
8120	3-1/2" thick	"	4.73	4.26	8.99
8140	4" thick	"	4.73	4.68	9.41
03730.10	**CONCRETE REPAIR**				
0090	Epoxy grout floor patch, 1/4" thick	SF	5.10	8.59	13.69
0100	Grout, epoxy, 2 component system	CF			420

03730.10 CONCRETE REPAIR, Cont'd...	UNIT	LABOR	MAT.	TOTAL
0110 Epoxy sand	BAG			28.00
0120 Epoxy modifier	GAL			180
0140 Epoxy gel grout	SF	51.00	4.18	55.18
0150 Injection valve, 1 way, threaded plastic	EA	10.25	11.50	21.75
0155 Grout crack seal, 2 component	CF	51.00	970	1,021
0160 Grout, non-shrink	"	51.00	99.00	150
0165 Concrete, epoxy modified				
0170 Sand mix	CF	20.50	160	181
0180 Gravel mix	"	19.00	110	129
0190 Concrete repair				
0195 Soffit repair				
0200 16" wide	LF	10.25	4.90	15.15
0210 18" wide	"	10.75	5.21	15.96
0220 24" wide	"	11.25	6.23	17.48
0230 30" wide	"	12.25	7.01	19.26
0240 32" wide	"	12.75	7.48	20.23
0245 Edge repair				
0250 2" spall	LF	12.75	2.33	15.08
0260 3" spall	"	13.50	2.33	15.83
0270 4" spall	"	13.75	2.48	16.23
0280 6" spall	"	14.25	2.57	16.82
0290 8" spall	"	15.00	2.72	17.72
0300 9" spall	"	17.00	2.80	19.80
0330 Crack repair, 1/8" crack	"	5.10	4.59	9.69
5000 Reinforcing steel repair				
5005 1 bar, 4 ft				
5010 #4 bar	LF	8.32	0.72	9.04
5012 #5 bar	"	8.32	0.97	9.29
5014 #6 bar	"	8.88	1.18	10.06
5016 #8 bar	"	8.88	2.14	11.02
5020 #9 bar	"	9.51	2.74	12.25
5030 #11 bar	"	9.51	4.28	13.79
7010 Form fabric, nylon				
7020 18" diameter	LF			18.50
7030 20" diameter	"			18.75
7040 24" diameter	"			31.00
7050 30" diameter	"			31.75
7060 36" diameter	"			36.50
7100 Pile repairs				
7105 Polyethylene wrap				
7108 30 mil thick				
7110 60" wide	SF	17.00	20.00	37.00
7120 72" wide	"	20.50	22.00	42.50
7125 60 mil thick				
7130 60" wide	SF	17.00	23.75	40.75
7140 80" wide	"	23.25	27.50	50.75
8010 Pile spall, average repair 3'				
8020 18" x 18"	EA	42.50	62.00	105
8030 20" x 20"	"	51.00	83.00	134

		UNIT	COST
03999.10	**CONCRETE**		
1000	FOOTINGS (Incl. exc. with steel)		
1010	By L.F - Continuous 24" x 12" (3 # 4 Rods)	LF	34.25
1020	36" x 12" (4 # 4 Rods)	"	43.50
1030	20" x 10" (2 # 5 Rods)	"	29.25
1040	16" x 8" (2 # 4 Rods)	"	25.00
1050	Pad 24" x 24" x 12" (4 # 5 E.W.)	EA	120
1060	36" x 36" x 14" (6 # 5 E.W.)	"	240
1070	48" x 48" x 16" (8 # 5 E.W.)	"	430
1100	By C.Y. - Continuous 24" x 12" (3 # 4 Rods)	CY	430
1110	36" x 12" (4 # 4 Rods)	"	450
1120	20" x 10" (2 # 5 Rods)	"	480
1130	16" x 8" (2 # 4 Rods)	"	580
1140	Pad 24" x 24" x 12" (4 # 5 E.W.)	"	720
1150	36" x 36" x 14" (6 # 5 E.W.)	"	680
1160	48" x 48" x 16" (8 # 5 E.W.)	"	650
1200	WALLS (#5 Rods 12" O.C. - 1 Face)		
1210	By S.F. - 8" Wall (# 5 12" O.C. E.W.)	SF	24.75
1220	12" Wall (# 5 12" O.C. E.W.)	"	26.75
1230	16" Wall (# 5 12" O.C. E.W.)	"	26.75
1300	Add for Steel - 2 Faces - 12" Wall	"	3.17
1310	Add for Pilastered Wall - 24" O.C.	"	1.51
1320	Add for Retaining or Battered Type	"	4.48
1330	Add for Curved Walls	"	8.31
1340	By C.Y. - 8" Wall (# 5 12" O.C. E.W.)	CY	800
1350	12" Wall (# 5 12" O.C. E.W.)	"	660
1360	16" Wall (# 5 12" O.C. E.W.)	"	600
1400	Add for Steel - 2 Faces - 12" Wall	"	88.25
1410	Add for Pilastered Wall - 24" O.C.	"	38.75
1420	Add for Retaining or Battered Type	"	121
1430	Add for Curved Walls	"	225
1500	COLUMNS		
1510	By L.F. - Square Cornered 8" x 8" (4 # 8 Rods)	LF	56.50
1520	12" x 12" (6 # 8 Rods)	"	95.75
1530	16" x 16" (6 # 10 Rods)	"	125
1540	20" x 20" (8 # 20 Rods)	"	178
1550	24" x 24" (10 # 11 Rods)	"	215
1600	Round 8" (4 # 8 Rods)	"	41.50
1610	12" (6 # 8 Rods)	"	64.75
1620	16" (6 # 10 Rods)	"	88.50
1630	20" (8 # 20 Rods)	"	124
1640	24" (10 # 11 Rods)	"	179
1700	By C.Y. - Sq Cornered 8" x 8" (4 # 8 Rods)	CY	3,280
1710	12" x 12" (6 # 8 Rods)	"	3,030
1720	16" x 16" (6 # 10 Rods)	"	2,150
1730	20" x 20" (8 # 20 Rods)	"	1,910
1740	24" x 24" (10 # 11 Rods)	"	1,370
1800	Round 8" (4 # 8 Rods)	"	3,030
1810	12" (6 # 8 Rods)	"	2,010
1820	16" (6 # 10 Rods)	"	1,850
1830	20" (8 # 20 Rods)	"	1,690
1840	24" (10 # 11 Rods)	"	1,610
1900	BEAMS		

		UNIT	COST
03999.10	**CONCRETE, Cont'd...**		
1910	By L.F. - Spandrel 12" x 48" (33 # rebar)	LF	215
1920	12" x 42" (26 # Rein. Steel)	"	195
1930	12" x 36" (21 # Rein. Steel)	"	156
1940	12" x 30" (15 # Rein. Steel)	"	148
1950	8" x 48" (26 # Rein. Steel)	"	194
1960	8" x 42" (21 # Rein. Steel)	"	175
1970	8" x 36" (16 # Rein. Steel)	"	158
2000	Interior 16" x 30" (24 # rebar)	"	126
2010	16" x 24" (20 # Rein. Steel)	"	125
2020	12" x 30" (17 # Rein. Steel)	"	146
2030	12" x 24" (14 # Rein. Steel)	"	105
2040	12" x 16" (12 # Rein. Steel)	"	88.50
2050	8" x 24" (13 # Rein. Steel)	"	105
2060	8" x 16" (10 # Rein. Steel)	"	78.50
2100	By C.Y. - Spandrel 12" x 48" (33 # rebar)	CY	1,290
2110	12" x 42" (26 # Rein. Steel)	"	1,370
2120	12" x 36" (21 # Rein. Steel)	"	1,450
2130	12" x 30" (15 # Rein. Steel)	"	1,530
2140	8" x 48" (26 # Rein. Steel)	"	1,530
2150	8" x 42" (21 # Rein. Steel)	"	1,670
2160	8" x 36" (16 # Rein. Steel)	"	1,740
2200	Interior 16" x 30" (24 # rebar)	"	850
2210	16" x 24" (20 # Rein. Steel)	"	1,250
2220	12" x 30" (17 # Rein. Steel)	"	1,370
2230	12" x 24" (14 # Rein. Steel)	"	1,340
2240	12" x 16" (12 # Rein. Steel)	"	1,520
2250	8" x 24" (13 # Rein. Steel)	"	1,690
2260	8" x 16" (10 # Rein. Steel)	"	1,770
2300	SLABS (With Reinf. Steel)		
2310	By S.F. - Solid		
2320	4" Thick	SF	15.24
2330	5" Thick	"	17.85
2340	6" Thick	"	19.69
2350	7" Thick	"	22.00
2360	8" Thick	"	21.07
2400	Deduct for Post Tensioned Slabs	"	1.31
2410	Pan	"	
2420	Joist -20" Pan - 10" x 2"	"	20.07
2430	12" x 2"	"	22.75
2440	30" Pan - 10" x 2-1/2"	"	20.25
2450	12" x 2-1/2"	"	22.75
2500	Dome -19" x 19" - 10" x 2"	"	22.25
2510	12" x 2"	"	23.00
2520	30" x 30" - 10" x 2-1/2"	"	19.50
2530	12" x 2-1/2"	"	21.75
2600	By C.Y. - Solid		
2610	4" Thick	CY	1,170
2620	5" Thick	"	1,110
2630	6" Thick	"	1,080
2640	8" Thick	"	1,040
2700	COMBINED COLUMNS, BEAMS AND SLABS		
2710	20' Span	SF	30.75

		UNIT	COST
03999.10	**CONCRETE, Cont'd...**		
2720	30' Span	SF	32.50
2730	40' Span	"	34.50
2740	50' Span	"	37.25
2750	60' Span	"	63.50
2800	STAIRS (Including Landing)		
2810	By EA. - 4' Wide 10' Floor Heights 16 Risers	EA	270
2820	5' Wide 10' Floor Heights 16 Risers	"	300
2830	6' Wide 10' Floor Heights 16 Risers	"	380
2840	By C.Y. - 4' Wide 10' Floor Heights 16 Risers	CY	1,690
2850	5' Wide 10' Floor Heights 16 Risers	"	1,740
2860	6' Wide 10' Floor Heights 16 Risers	"	1,710
2900	SLABS ON GROUND		
2910	4" Concrete Slab		
2920	(6 6/10 - 10 Mesh - 5 1/2 Sack Concrete,		
2930	Trowel Finished, Cured & Truck Chuted)	SF	5.19
2940	Add per Inch of Concrete	"	0.87
2950	Add per Sack of Cement	"	0.23
2960	Deduct for Float Finish	"	0.11
2970	Deduct for Brush or Broom Finish	"	0.09
3000	Add for Runway and Buggied Concrete	"	0.34
3010	Add for Vapor Barrier (4 mil)	"	0.23
3020	Add for Sub - Floor Fill (4" sand/gravel)	"	0.73
3030	Add for Change to 6 6/8 - 8 Mesh	"	0.12
3040	Add for Change to 6 6/6 - 6 Mesh	"	0.23
3050	Add for Sloped Slab	"	0.32
3060	Add for Edge Strip (sidewalk area)	"	0.34
3100	Add for ½" Expansion Joint (20' O.C.)	"	0.11
3110	Add for Control Joints (keyed/ dep.)	"	0.26
3120	Add for Control Joints (joint filled)	"	0.28
3130	Add for Control Joints - Saw Cut (20' O.C.)	"	0.48
3140	Add for Floor Hardener (1 coat)	"	0.22
3150	Add for Exposed Aggregate - Washed Added	"	0.65
3200	Retarding Added	"	0.62
3210	Seeding Added	"	0.65
3220	Add for Light Weight Aggregates	"	0.65
3230	Add for Heavy Weight Aggregates	"	0.45
3240	Add for Winter Production Loss/Cost	"	0.57
3300	TOPPING SLABS		
3310	2" Concrete	SF	4.43
3320	3" Concrete	"	5.52
3330	No Mesh or Hoisting		
3400	PADS & PLATFORMS (Including Form Work & Reinforcing)		
3410	4"	SF	10.81
3420	6"	"	13.64
3500	PRECAST CONCRETE ITEMS		
3510	Curbs 6" x 10" x 8"	LF	18.36
3520	Sills & Stools 6"	"	41.50
3530	Splash Blocks 3" x 16"	EA	101
3600	MISCELLANEOUS ADDITIONS TO ABOVE CONCRETE		
3610	Abrasives - Carborundum - Grits	SF	1.42
3620	Strips	LF	4.67
3630	Bushhammer - Green Concrete	SF	3.14

		UNIT	COST
03999.10	**CONCRETE, Cont'd...**		
3640	Cured Concrete	SF	4.21
3650	Chamfers - Plastic 3/4"	LF	1.65
3660	Wood 3/4"	"	0.98
3670	Metal 3/4"	"	1.80
3680	Colors Dust On	SF	1.32
3690	Integral (Top 1")	"	2.23
3700	Control Joints - Asphalt 1/2" x 4"	LF	1.40
3710	1/2" x 6"	"	1.62
3720	PolyFoam 1/2" x 4"	"	1.75
3730	Dovetail Slots 22 Ga	"	2.44
3740	24 Ga	"	1.56
3800	Hardeners Acrylic and Urethane	SF	0.47
3810	Epoxy	"	0.59
3820	Joint Sealers Epoxy	LF	5.60
3830	Rubber Asphalt	"	2.96
3840	Moisture Proofing - Polyethylene 4 mil	SF	0.30
3850	6 mil	"	0.36
3900	Non-Shrink Grouts - Non - Metallic	CF	64.75
3910	Aluminum Oxide	"	47.00
3920	Iron Oxide	"	69.25
4000	Reglets Flashing	LF	4.31
4010	Sand Blast - Light	SF	2.23
4020	Heavy	"	4.33
4030	Shelf Angle Inserts 5/8"	LF	15.60
4040	Stair Nosings Steel - Galvanized	"	11.29
4050	Tongue & Groove Joint Forms - Asphalt 5-1/2"	"	3.39
4060	Wood 5-1/2"	"	2.78
4070	Metal 5-1/2"	"	3.52
4100	Treads - Extruded Aluminum	"	14.87
4110	Cast Iron	"	17.19
4120	Water Stops Center Bulb - Rubber - 6"	"	17.55
4130	9"	"	33.00
4140	Polyethylene - 6"	"	4.84
4150	9"	"	4.90

DCD

Design Cost Data™

TABLE OF CONTENTS   PAGE

		UNIT	LABOR	MAT.	TOTAL
04100.10	**MASONRY GROUT**				
0100	Grout, non-shrink, non-metallic, trowelable	CF	1.90	5.94	7.84
2110	Grout door frame, hollow metal				
2120	Single	EA	71.00	14.50	85.50
2140	Double	"	75.00	20.50	95.50
2980	Grout-filled concrete block (CMU)				
3000	4" wide	SF	2.37	0.43	2.80
3020	6" wide	"	2.59	1.12	3.71
3040	8" wide	"	2.85	1.65	4.50
3060	12" wide	"	3.00	2.71	5.71
3070	Grout-filled individual CMU cells				
3090	4" wide	LF	1.42	0.36	1.78
3100	6" wide	"	1.42	0.48	1.90
3120	8" wide	"	1.42	0.64	2.06
3140	10" wide	"	1.62	0.80	2.42
3160	12" wide	"	1.62	0.97	2.59
4000	Bond beams or lintels, 8" deep				
4010	6" thick	LF	2.34	0.97	3.31
4020	8" thick	"	2.58	1.28	3.86
4040	10" thick	"	2.86	1.61	4.47
4060	12" thick	"	3.22	1.93	5.15
5000	Cavity walls				
5020	2" thick	SF	3.44	1.07	4.51
5040	3" thick	"	3.44	1.61	5.05
5060	4" thick	"	3.68	2.14	5.82
5080	6" thick	"	4.30	3.21	7.51
04150.10	**MASONRY ACCESSORIES**				
0200	Foundation vents	EA	24.75	26.25	51.00
1010	Bar reinforcing				
1015	Horizontal				
1020	#3 - #4	LB	2.48	0.60	3.08
1030	#5 - #6	"	2.07	0.60	2.67
1035	Vertical				
1040	#3 - #4	LB	3.11	0.60	3.71
1050	#5 - #6	"	2.48	0.60	3.08
1100	Horizontal joint reinforcing				
1105	Truss type				
1110	4" wide, 6" wall	LF	0.24	0.20	0.44
1120	6" wide, 8" wall	"	0.25	0.20	0.45
1130	8" wide, 10" wall	"	0.27	0.25	0.52
1140	10" wide, 12" wall	"	0.28	0.25	0.53
1150	12" wide, 14" wall	"	0.29	0.30	0.59
1155	Ladder type				
1160	4" wide, 6" wall	LF	0.24	0.15	0.39
1170	6" wide, 8" wall	"	0.25	0.17	0.42
1180	8" wide, 10" wall	"	0.27	0.18	0.45
1190	10" wide, 12" wall	"	0.27	0.22	0.49
2000	Rectangular wall ties				
2005	3/16" dia., galvanized				
2010	2" x 6"	EA	1.03	0.38	1.41
2020	2" x 8"	"	1.03	0.40	1.43
2040	2" x 10"	"	1.03	0.47	1.50

		UNIT	LABOR	MAT.	TOTAL
04150.10	**MASONRY ACCESSORIES, Cont'd...**				
2050	2" x 12"	EA	1.03	0.53	1.56
2060	4" x 6"	"	1.24	0.44	1.68
2070	4" x 8"	"	1.24	0.49	1.73
2080	4" x 10"	"	1.24	0.63	1.87
2090	4" x 12"	"	1.24	0.73	1.97
2095	1/4" dia., galvanized				
2100	2" x 6"	EA	1.03	0.71	1.74
2110	2" x 8"	"	1.03	0.80	1.83
2120	2" x 10"	"	1.03	0.91	1.94
2130	2" x 12"	"	1.03	1.04	2.07
2140	4" x 6"	"	1.24	0.82	2.06
2150	4" x 8"	"	1.24	0.91	2.15
2160	4" x 10"	"	1.24	1.04	2.28
2170	4" x 12"	"	1.24	1.08	2.32
2200	Z-type wall ties, galvanized				
2215	6" long				
2220	1/8" dia.	EA	1.03	0.34	1.37
2230	3/16" dia.	"	1.03	0.36	1.39
2240	1/4" dia.	"	1.03	0.38	1.41
2245	8" long				
2250	1/8" dia.	EA	1.03	0.36	1.39
2260	3/16" dia.	"	1.03	0.38	1.41
2270	1/4" dia.	"	1.03	0.40	1.43
2275	10" long				
2280	1/8" dia.	EA	1.03	0.38	1.41
2290	3/16" dia.	"	1.03	0.44	1.47
2300	1/4" dia.	"	1.03	0.49	1.52
3000	Dovetail anchor slots				
3015	Galvanized steel, filled				
3020	24 ga.	LF	1.55	1.17	2.72
3040	20 ga.	"	1.55	2.46	4.01
3060	16 oz. copper, foam filled	"	1.55	3.53	5.08
3100	Dovetail anchors				
3115	16 ga.				
3120	3-1/2" long	EA	1.03	0.39	1.42
3140	5-1/2" long	"	1.03	0.48	1.51
3150	12 ga.				
3160	3-1/2" long	EA	1.03	0.52	1.55
3180	5-1/2" long	"	1.03	0.86	1.89
3200	Dovetail, triangular galvanized ties, 12 ga.				
3220	3" x 3"	EA	1.03	0.88	1.91
3240	5" x 5"	"	1.03	0.95	1.98
3260	7" x 7"	"	1.03	1.07	2.10
3280	7" x 9"	"	1.03	1.14	2.17
3400	Brick anchors				
3420	Corrugated, 3-1/2" long				
3440	16 ga.	EA	1.03	0.57	1.60
3460	12 ga.	"	1.03	0.66	1.69
3500	Non-corrugated, 3-1/2" long				
3520	16 ga.	EA	1.03	0.47	1.50
3540	12 ga.	"	1.03	0.85	1.88
3580	Cavity wall anchors, corrugated, galvanized				

		UNIT	LABOR	MAT.	TOTAL
04150.10	**MASONRY ACCESSORIES, Cont'd...**				
3600	5" long				
3620	16 ga.	EA	1.03	0.95	1.98
3640	12 ga.	"	1.03	1.43	2.46
3660	7" long				
3680	28 ga.	EA	1.03	1.05	2.08
3700	24 ga.	"	1.03	1.33	2.36
3720	22 ga.	"	1.03	1.36	2.39
3740	16 ga.	"	1.03	1.55	2.58
3800	Mesh ties, 16 ga., 3" wide				
3820	8" long	EA	1.03	1.28	2.31
3840	12" long	"	1.03	1.43	2.46
3860	20" long	"	1.03	1.96	2.99
3900	24" long	"	1.03	2.16	3.19
04150.20	**MASONRY CONTROL JOINTS**				
1000	Control joint, cross shaped PVC	LF	1.55	2.38	3.93
1010	Closed cell joint filler				
1020	1/2"	LF	1.55	0.41	1.96
1040	3/4"	"	1.55	0.85	2.40
1070	Rubber, for				
1080	4" wall	LF	1.55	2.75	4.30
1090	6" wall	"	1.63	3.40	5.03
1100	8" wall	"	1.72	4.10	5.82
1110	PVC, for				
1120	4" wall	LF	1.55	1.43	2.98
1140	6" wall	"	1.63	2.41	4.04
1160	8" wall	"	1.72	3.65	5.37
04150.50	**MASONRY FLASHING**				
0080	Through-wall flashing				
1000	5 oz. coated copper	SF	5.18	4.18	9.36
1020	0.030" elastomeric	"	4.14	1.32	5.46
04210.10	**BRICK MASONRY**				
0100	Standard size brick, running bond				
1000	Face brick, red (6.4/sf)				
1020	Veneer	SF	10.25	5.66	15.91
1030	Cavity wall	"	8.88	5.66	14.54
1040	9" solid wall	"	17.75	11.25	29.00
1200	Common brick (6.4/sf)				
1210	Select common for veneers	SF	10.25	3.69	13.94
1215	Back-up				
1220	4" thick	SF	7.77	3.32	11.09
1230	8" thick	"	12.50	6.65	19.15
1235	Firewall				
1240	12" thick	SF	20.75	10.75	31.50
1250	16" thick	"	28.25	14.25	42.50
1300	Glazed brick (7.4/sf)				
1310	Veneer	SF	11.25	15.25	26.50
1400	Buff or gray face brick (6.4/sf)				
1410	Veneer	SF	10.25	6.58	16.83
1420	Cavity wall	"	8.88	6.58	15.46
1500	Jumbo or oversize brick (3/sf)				
1510	4" veneer	SF	6.22	4.76	10.98

DIVISION # 04 MASONRY

		UNIT	LABOR	MAT.	TOTAL
04210.10	**BRICK MASONRY, Cont'd...**				
1530	4" back-up	SF	5.18	4.76	9.94
1540	8" back-up	"	8.88	5.52	14.40
1550	12" firewall	"	15.50	7.42	22.92
1560	16" firewall	"	20.75	10.50	31.25
1600	Norman brick, red face (4.5/sf)				
1620	4" veneer	SF	7.77	7.88	15.65
1640	Cavity wall	"	6.91	7.88	14.79
3000	Chimney, standard brick, including flue				
3020	16" x 16"	LF	62.00	32.00	94.00
3040	16" x 20"	"	62.00	54.00	116
3060	16" x 24"	"	62.00	58.00	120
3080	20" x 20"	"	78.00	45.00	123
3100	20" x 24"	"	78.00	61.00	139
3120	20" x 32"	"	89.00	68.00	157
4000	Window sill, face brick on edge	"	15.50	3.59	19.09
04210.20	**STRUCTURAL TILE**				
5000	Structural glazed tile				
5010	6T series, 5-1/2" x 12"				
5020	Glazed on one side				
5040	2" thick	SF	6.22	11.00	17.22
5060	4" thick	"	6.22	13.25	19.47
5080	6" thick	"	6.91	20.75	27.66
5100	8" thick	"	7.77	25.50	33.27
5200	Glazed on two sides				
5220	4" thick	SF	7.77	19.25	27.02
5240	6" thick	"	8.88	26.50	35.38
5500	Special shapes				
5510	Group 1	SF	12.50	11.25	23.75
5520	Group 2	"	12.50	14.25	26.75
5530	Group 3	"	12.50	18.75	31.25
5540	Group 4	"	12.50	37.75	50.25
5550	Group 5	"	12.50	46.00	58.50
5600	Fire rated				
5620	4" thick, 1 hr rating	SF	6.22	18.00	24.22
5640	6" thick, 2 hr rating	"	6.91	25.25	32.16
6000	8W series, 8" x 16"				
6010	Glazed on one side				
6020	2" thick	SF	4.14	12.50	16.64
6040	4" thick	"	4.14	13.25	17.39
6060	6" thick	"	4.78	22.00	26.78
6080	8" thick	"	4.78	24.00	28.78
6100	Glazed on two sides				
6120	4" thick	SF	5.18	21.00	26.18
6140	6" thick	"	6.22	29.25	35.47
6160	8" thick	"	6.22	35.50	41.72
6200	Special shapes				
6220	Group 1	SF	8.88	18.75	27.63
6230	Group 2	"	8.88	23.00	31.88
6240	Group 3	"	8.88	25.25	34.13
6250	Group 4	"	8.88	42.00	50.88
6260	Group 5	"	8.88	53.00	61.88

		UNIT	LABOR	MAT.	TOTAL
04210.20	**STRUCTURAL TILE, Cont'd...**				
6270	Fire rated				
6290	4" thick, 1 hr rating	SF	8.88	31.50	40.38
6300	6" thick, 2 hr rating	"	8.88	43.25	52.13
04210.60	**PAVERS, MASONRY**				
4000	Brick walk laid on sand, sand joints				
4020	Laid flat (4.5 per sf)	SF	6.91	4.16	11.07
4040	Laid on edge (7.2 per sf)	"	10.25	6.66	16.91
5000	Precast concrete patio blocks				
5005	2" thick				
5010	Natural	SF	2.07	3.58	5.65
5020	Colors	"	2.07	4.53	6.60
5080	Exposed aggregates, local aggregate				
5100	Natural	SF	2.07	10.00	12.07
5120	Colors	"	2.07	10.00	12.07
5130	Granite or limestone aggregate	"	2.07	10.00	12.07
5140	White tumblestone aggregate	"	2.07	10.75	12.82
6000	Stone pavers, set in mortar				
6005	Bluestone				
6008	1" thick				
6010	Irregular	SF	15.50	10.00	25.50
6020	Snapped rectangular	"	12.50	14.75	27.25
6060	1-1/2" thick, random rectangular	"	15.50	18.00	33.50
6070	2" thick, random rectangular	"	17.75	20.50	38.25
6090	Slate				
6100	Natural cleft				
6110	Irregular, 3/4" thick	SF	17.75	10.75	28.50
6115	Random rectangular				
6120	1-1/4" thick	SF	15.50	23.25	38.75
6130	1-1/2" thick	"	17.25	26.25	43.50
7000	Granite blocks				
7010	3" thick, 3" to 6" wide				
7020	4" to 12" long	SF	20.75	13.25	34.00
7030	6" to 15" long	"	17.75	8.67	26.42
9000	Flagstone pavers				
9010	Random sizes, 1-4 sf, tumbled cobble	SF	12.50	17.00	29.50
9020	Tumbled patio	"	12.50	18.25	30.75
9040	Saw-cut Flagstone Tiles				
9060	12"x12", assorted colors	SF	8.88	15.75	24.63
9080	18"x18", assorted colors	"	7.77	15.75	23.52
9100	24"x24", assorted colors	"	6.91	15.75	22.66
9800	Crushed stone, white marble, 3" thick	"	1.02	1.88	2.90
04220.10	**CONCRETE MASONRY UNITS**				
0110	Hollow, load bearing				
0120	4"	SF	4.60	1.62	6.22
0140	6"	"	4.78	2.38	7.16
0160	8"	"	5.18	2.73	7.91
0180	10"	"	5.65	3.77	9.42
0190	12"	"	6.22	4.34	10.56
0280	Solid, load bearing				
0300	4"	SF	4.60	2.55	7.15
0320	6"	"	4.78	2.86	7.64

04220.10	CONCRETE MASONRY UNITS, Cont'd...	UNIT	LABOR	MAT.	TOTAL
0340	8"	SF	5.18	3.91	9.09
0360	10"	"	5.65	4.16	9.81
0380	12"	"	6.22	6.19	12.41
0480	Back-up block, 8" x 16"				
0500	2"	SF	3.55	1.70	5.25
0540	4"	"	3.65	1.78	5.43
0560	6"	"	3.88	2.60	6.48
0580	8"	"	4.14	2.99	7.13
0600	10"	"	4.44	4.13	8.57
0620	12"	"	4.78	4.75	9.53
0980	Foundation wall, 8" x 16"				
1000	6"	SF	4.44	2.60	7.04
1030	8"	"	4.78	2.99	7.77
1040	10"	"	5.18	4.13	9.31
1050	12"	"	5.65	4.76	10.41
1055	Solid				
1060	6"	SF	4.78	3.15	7.93
1070	8"	"	5.18	4.30	9.48
1080	10"	"	5.65	4.57	10.22
1100	12"	"	6.22	6.79	13.01
1480	Exterior, styrofoam inserts, std weight, 8" x 16"				
1500	6"	SF	4.78	4.57	9.35
1530	8"	"	5.18	4.93	10.11
1540	10"	"	5.65	6.40	12.05
1550	12"	"	6.22	8.77	14.99
1580	Lightweight				
1600	6"	SF	4.78	5.09	9.87
1660	8"	"	5.18	5.73	10.91
1680	10"	"	5.65	6.08	11.73
1700	12"	"	6.22	8.04	14.26
1980	Acoustical slotted block				
2000	4"	SF	5.65	5.31	10.96
2020	6"	"	5.65	5.56	11.21
2040	8"	"	6.22	6.94	13.16
2050	Filled cavities				
2060	4"	SF	6.91	5.69	12.60
2070	6"	"	7.31	6.55	13.86
2080	8"	"	7.77	8.40	16.17
4000	Hollow, split face				
4020	4"	SF	4.60	3.64	8.24
4030	6"	"	4.78	4.21	8.99
4040	8"	"	5.18	4.42	9.60
4080	10"	"	5.65	4.95	10.60
4100	12"	"	6.22	5.28	11.50
4480	Split rib profile				
4500	4"	SF	5.65	4.42	10.07
4520	6"	"	5.65	5.13	10.78
4540	8"	"	6.22	5.58	11.80
4560	10"	"	6.22	6.12	12.34
4580	12"	"	6.22	6.63	12.85
4980	High strength block, 3500 psi				
5000	2"	SF	4.60	1.71	6.31

		UNIT	LABOR	MAT.	TOTAL
04220.10	**CONCRETE MASONRY UNITS, Cont'd...**				
5020	4"	SF	4.78	2.14	6.92
5030	6"	"	4.78	2.56	7.34
5040	8"	"	5.18	2.90	8.08
5050	10"	"	5.65	3.38	9.03
5060	12"	"	6.22	4.00	10.22
5500	Solar screen concrete block				
5505	4" thick				
5510	6" x 6"	SF	13.75	4.29	18.04
5520	8" x 8"	"	12.50	5.12	17.62
5530	12" x 12"	"	9.56	5.24	14.80
5540	8" thick				
5550	8" x 16"	SF	8.88	5.24	14.12
7000	Glazed block				
7020	Cove base, glazed 1 side, 2"	LF	6.91	11.00	17.91
7030	4"	"	6.91	11.25	18.16
7040	6"	"	7.77	11.50	19.27
7050	8"	"	7.77	12.50	20.27
7055	Single face				
7060	2"	SF	5.18	11.50	16.68
7080	4"	"	5.18	14.25	19.43
7090	6"	"	5.65	15.25	20.90
7100	8"	"	6.22	16.00	22.22
7105	10"	"	6.91	18.00	24.91
7110	12"	"	7.31	19.25	26.56
7115	Double face				
7120	4"	SF	6.54	17.25	23.79
7140	6"	"	6.91	20.25	27.16
7160	8"	"	7.77	21.25	29.02
7180	Corner or bullnose				
7200	2"	EA	7.77	18.25	26.02
7240	4"	"	8.88	23.50	32.38
7260	6"	"	8.88	28.75	37.63
7280	8"	"	10.25	31.25	41.50
7290	10"	"	11.25	34.00	45.25
7300	12"	"	12.50	36.50	49.00
9500	Gypsum unit masonry				
9510	Partition blocks (12"x30")				
9515	Solid				
9520	2"	SF	2.48	1.41	3.89
9525	Hollow				
9530	3"	SF	2.48	1.42	3.90
9540	4"	"	2.59	1.63	4.22
9550	6"	"	2.82	1.74	4.56
9900	Vertical reinforcing				
9920	4' o.c., add 5% to labor				
9940	2'8" o.c., add 15% to labor				
9960	Interior partitions, add 10% to labor				
04220.90	**BOND BEAMS & LINTELS**				
0980	Bond beam, no grout or reinforcement				
0990	8" x 16" x				
1000	4" thick	LF	4.78	1.85	6.63

	04220.90 BOND BEAMS & LINTELS, Cont'd...	UNIT	LABOR	MAT.	TOTAL
1040	6" thick	LF	4.97	2.83	7.80
1060	8" thick	"	5.18	3.24	8.42
1080	10" thick	"	5.40	4.01	9.41
1100	12" thick	"	5.65	4.56	10.21
6000	Beam lintel, no grout or reinforcement				
6010	8" x 16" x				
6020	10" thick	LF	6.22	8.85	15.07
6040	12" thick	"	6.91	9.42	16.33
6080	Precast masonry lintel				
7000	6 lf, 8" high x				
7020	4" thick	LF	10.25	7.80	18.05
7040	6" thick	"	10.25	9.96	20.21
7060	8" thick	"	11.25	11.25	22.50
7080	10" thick	"	11.25	13.50	24.75
7090	10 lf, 8" high x				
7100	4" thick	LF	6.22	9.80	16.02
7120	6" thick	"	6.22	12.00	18.22
7140	8" thick	"	6.91	13.50	20.41
7160	10" thick	"	6.91	18.25	25.16
8000	Steel angles and plates				
8010	Minimum	LB	0.88	1.33	2.21
8020	Maximum	"	1.55	1.95	3.50
8200	Various size angle lintels				
8205	1/4" stock				
8210	3" x 3"	LF	3.88	6.85	10.73
8220	3" x 3-1/2"	"	3.88	7.54	11.42
8225	3/8" stock				
8230	3" x 4"	LF	3.88	12.00	15.88
8240	3-1/2" x 4"	"	3.88	12.50	16.38
8250	4" x 4"	"	3.88	13.75	17.63
8260	5" x 3-1/2"	"	3.88	14.50	18.38
8262	6" x 3-1/2"	"	3.88	16.25	20.13
8265	1/2" stock				
8280	6" x 4"	LF	3.88	18.00	21.88
04240.10	**CLAY TILE**				
0100	Hollow clay tile, for back-up, 12" x 12"				
1000	Scored face				
1010	Load bearing				
1020	4" thick	SF	4.44	7.22	11.66
1040	6" thick	"	4.60	8.42	13.02
1060	8" thick	"	4.78	10.50	15.28
1080	10" thick	"	4.97	13.00	17.97
1100	12" thick	"	5.18	22.00	27.18
2000	Non-load bearing				
2020	3" thick	SF	4.28	5.85	10.13
2040	4" thick	"	4.44	6.77	11.21
2060	6" thick	"	4.60	7.84	12.44
2080	8" thick	"	4.78	9.99	14.77
2100	12" thick	"	5.18	17.75	22.93
4100	Partition, 12" x 12"				
4150	In walls				

		UNIT	LABOR	MAT.	TOTAL
04240.10	**CLAY TILE, Cont'd...**				
4201	3" thick	SF	5.18	5.85	11.03
4210	4" thick	"	5.18	6.78	11.96
4220	6" thick	"	5.40	7.48	12.88
4230	8" thick	"	5.65	9.81	15.46
4240	10" thick	"	5.92	11.75	17.67
4250	12" thick	"	6.22	17.00	23.22
4300	Clay tile floors				
4320	4" thick	SF	3.45	6.77	10.22
4330	6" thick	"	3.65	8.42	12.07
4340	8" thick	"	3.88	10.50	14.38
4350	10" thick	"	4.14	13.00	17.14
4360	12" thick	"	4.44	19.25	23.69
6000	Terra cotta				
6020	Coping, 10" or 12" wide, 3" thick	LF	12.50	16.00	28.50
04270.10	**GLASS BLOCK**				
1000	Glass block, 4" thick				
1040	6" x 6"	SF	20.75	38.25	59.00
1060	8" x 8"	"	15.50	24.25	39.75
1080	12" x 12"	"	12.50	30.75	43.25
8980	Replacement glass blocks, 4" x 8" x 8"				
9100	Minimum	SF	62.00	17.75	79.75
9120	Maximum	"	120	30.50	151
04295.10	**PARGING / MASONRY PLASTER**				
0080	Parging				
0100	1/2" thick	SF	4.14	0.44	4.58
0200	3/4" thick	"	5.18	0.48	5.66
0300	1" thick	"	6.22	0.63	6.85
04400.10	**STONE**				
0160	Rubble stone				
0180	Walls set in mortar				
0200	8" thick	SF	15.50	17.50	33.00
0220	12" thick	"	24.75	21.00	45.75
0420	18" thick	"	31.00	28.00	59.00
0440	24" thick	"	41.50	35.00	76.50
0445	Dry set wall				
0450	8" thick	SF	10.25	19.50	29.75
0455	12" thick	"	15.50	22.25	37.75
0460	18" thick	"	20.75	30.50	51.25
0465	24" thick	"	24.75	37.25	62.00
0480	Cut stone				
0490	Imported marble				
0510	Facing panels				
0520	3/4" thick	SF	24.75	45.25	70.00
0530	1-1/2" thick	"	28.25	64.00	92.25
0540	2-1/4" thick	"	34.50	77.00	112
0600	Base				
0610	1" thick				
0620	4" high	LF	31.00	20.50	51.50
0640	6" high	"	31.00	24.75	55.75
0700	Columns, solid				
0720	Plain faced	CF	410	150	560

		UNIT	LABOR	MAT.	TOTAL
04400.10	**STONE, Cont'd...**				
0740	Fluted	CF	410	420	830
0780	Flooring, travertine, minimum	SF	9.56	20.75	30.31
0800	Average	"	12.50	27.75	40.25
0820	Maximum	"	13.75	51.00	64.75
1000	Domestic marble				
1020	Facing panels				
1040	7/8" thick	SF	24.75	42.50	67.25
1060	1-1/2" thick	"	28.25	64.00	92.25
1080	2-1/4" thick	"	34.50	77.00	112
1500	Stairs				
1510	12" treads	LF	31.00	38.00	69.00
1520	6" risers	"	20.75	28.25	49.00
1525	Thresholds, 7/8" thick, 3' long, 4" to 6" wide				
1530	Plain	EA	52.00	34.50	86.50
1540	Beveled	"	52.00	38.25	90.25
1545	Window sill				
1550	6" wide, 2" thick	LF	24.75	19.25	44.00
1555	Stools				
1560	5" wide, 7/8" thick	LF	24.75	25.75	50.50
1620	Limestone panels up to 12' x 5', smooth finish				
1630	2" thick	SF	11.50	29.50	41.00
1650	3" thick	"	11.50	34.50	46.00
1660	4" thick	"	11.50	49.25	60.75
1760	Miscellaneous limestone items				
1770	Steps, 14" wide, 6" deep	LF	41.50	63.00	105
1780	Coping, smooth finish	CF	20.75	88.00	109
1790	Sills, lintels, jambs, smooth finish	"	24.75	88.00	113
1800	Granite veneer facing panels, polished				
1810	7/8" thick				
1820	Black	SF	24.75	48.25	73.00
1840	Gray	"	24.75	38.00	62.75
1850	Base				
1860	4" high	LF	12.50	20.25	32.75
1870	6" high	"	13.75	24.50	38.25
1880	Curbing, straight, 6" x 16"	"	47.50	22.50	70.00
1890	Radius curbs, radius over 5'	"	63.00	27.50	90.50
1900	Ashlar veneer				
1905	4" thick, random	SF	24.75	34.00	58.75
1910	Pavers, 4" x 4" split				
1915	Gray	SF	12.50	33.25	45.75
1920	Pink	"	12.50	32.75	45.25
1930	Black	"	12.50	32.25	44.75
2000	Slate, panels				
2010	1" thick	SF	24.75	27.50	52.25
2020	2" thick	"	28.25	37.25	65.50
2030	Sills or stools				
2040	1" thick				
2060	6" wide	LF	24.75	12.75	37.50
2080	10" wide	"	27.00	20.75	47.75
2100	2" thick				
2120	6" wide	LF	28.25	21.00	49.25
2140	10" wide	"	31.00	34.75	65.75

		UNIT	COST
04999.10	**MASONRY**		
1000	BRICK MASONRY		
1100	Conventional (Modular Size)		
1110	Running Bond - 8" x 2-2/3" x 4"	SF	20.91
1120	Common Bond - 6 Course Header	"	24.43
1130	Stack Bond	"	21.26
1140	Dutch & English Bond - Every Other Course Header	"	31.25
1150	Every Course Header	"	34.75
1160	Flemish Bond - Every Other Course Header	"	23.66
1170	Every Course Header	"	28.00
1200	Add: If Scaffold Needed	"	1.05
1210	Add: For each 10' of Floor Hgt. or Floor (3%)	"	0.58
1220	Add: Piers and Corbels (15% to Labor)	"	3.00
1230	Add: Sills and Soldiers (20% to Labor)	"	2.86
1240	Add: Floor Brick (10% to Labor)	"	2.09
1300	Add: Weave & Herringbone Patterns (20% to Labor)	"	2.85
1310	Add: Stack Bond (8% to Labor)	"	1.65
1320	Add: Circular or Radius Work 20% to Labor)	"	4.37
1330	Add: Rock Faced & Slurried Face (10% to Labor)	"	2.09
1340	Add: Arches (75% to Labor)	"	6.42
1350	Add: For Winter Work (below 40°)	"	2.09
1400	Production Loss (10% to Labor)	"	2.09
1410	Enclosures - Wall Area Conventional	"	2.09
1420	Heat and Fuel - Wall Area Conventional	"	0.42
1500	Econo - 8" x 4" x 3"	"	17.30
1600	Panel - 8" x 8" x 4"	"	13.55
1700	Norman - 12" x 2-2/3" x 4"	"	17.33
1800	King Size - 10" x 2-5/8" x 4"	"	15.03
1900	Norwegian - 12" x 3-1/5" x 4"	"	14.48
2000	Saxon-Utility - 12" x 4" x 3"	"	13.58
2010	12" x 4" x 4"	"	13.62
2020	12" x 4" x 6"	"	17.26
2030	12" x 4" x 8"	"	20.57
2200	Adobe - 12" x 3" x 4"	"	23.51
2300	COATED BRICK (Ceramic)	"	28.00
2400	COMMON BRICK (Clay, Concrete and Sand Lime)	"	16.07
2500	FIRE BRICK -Light Duty	"	26.26
2510	Heavy Duty	"	33.25
2520	Deduct for Residential Work - All Above - 5%		
04999.20	**CONCRETE BLOCK**		
1000	CONVENTIONAL (Struck 2 Sides - Partitions)		
1010	12" x 8"x 16" Plain	SF	10.86
1020	Bond Beam (with Fill-Reinforcing)	"	14.37
1030	12" x 8"x 8" Half Block	"	13.10
1040	Double End - Header	"	11.13
1050	8" x 8" x 16" Plain	"	9.48
1060	Bond Beam (with Fill-Reinforcing)	"	11.64
1070	8" x 8" x 8" Half Block	"	9.28
1080	Double End - Header	"	9.86
1090	6 "x 8" x 16" Plain	"	8.85
2000	Bond Beam (with Fill-Reinforcing)	"	9.91
2010	Half Block	"	8.71

DIVISION # 04 MASONRY - QUICK ESTIMATING

		UNIT	COST
04999.20	**CONCRETE BLOCK, Cont'd...**		
2020	4" x 8" x 16" Plain	SF	8.56
2030	Half Block	"	7.46
2040	16" x 8" x 16" Plain	"	12.31
2050	Half Block	"	11.74
2060	14" x 8" x 16" Plain	"	11.69
2070	Half Block	"	11.10
2080	10" x 8" x 16" Plain	"	10.12
3000	Bond Beam (with Fill-Reinforcing)	"	11.67
3010	Half Block	"	11.35
3020	Deduct: Block Struck or Cleaned One Side	"	0.35
3030	Deduct: Block Not Struck or Cleaned Two Sides	"	0.57
3040	Deduct: Clean One Side Only	"	0.25
3050	Deduct: Lt. Wt. Block (Labor Only)	"	0.23
4000	Add: Jamb and Sash Block	"	0.60
4010	Add: Bullnose Block	"	0.89
4020	Add: Pilaster, Pier, Pedestal Work	"	0.98
4030	Add: Stack Bond Work	"	0.36
4040	Add: Radius or Circular Work	"	1.83
5000	Add: If Scaffold Needed	"	1.00
5010	Add: Winter Production Cost (10% Labor)	"	0.59
5020	Add: Winter Enclosing - Wall Area	"	0.59
5030	Add: Winter Heating - Wall Area	"	0.54
5040	Add: Core & Beam Filling - See 0411 (Page 4A-18)		
5050	Add: Insulation - See 0412 (Page 4A-18)		
6000	Deduct for Residential Work - 5%		
8010	SCREEN WALL - 4" x 12" x 12"	SF	8.52
8020	BURNISHED - 12" x 8" x 16"	"	16.04
8030	8" x 8" x 16"	"	14.88
8040	6" x 8" x 16"	"	13.58
8050	4" x 8" x 16"	"	12.75
8060	2" x 8" x 16"	"	11.48
8070	Add for Shapes	"	3.46
8080	Add for 2 Faced Finish	"	4.98
8090	Add for Scored Finish - 1 Face	"	1.21
8100	PREFACED UNITS -12" x 8" x 16" Stretcher	"	22.42
8110	(Ceramic Glazed) 12" x 8" x 16" Glazed 2 Face	"	30.00
8120	8" x 8" x 16" Stretcher	"	20.84
8130	Glazed 2 Face	"	28.25
8140	6" x 8" x 16" Stretcher	"	18.53
8150	Glazed 2 Face	"	28.25
8160	4" x 8" x 16" Stretcher	"	19.66
8170	Glazed 2 Face	"	27.15
8180	2" x 8" x 16" Stretcher	"	19.52
8190	4" x 16" x 16" Stretcher	"	50.00
8200	Add for Base, Caps, Jambs, Headers, Lintels	"	5.60
8210	Add for Scored Block	"	1.65
04999.30	**CLAY BACKING AND PARTITION TILE**		
0010	3" x 12" x 12"	SF	7.79
0020	4" x 12" x 12"	"	8.34
0030	6" x 12" x 12"	"	9.94
0040	8" x 12" x 12"	"	10.88

		UNIT	COST
04999.40	**CLAY FACING TILE (GLAZED)**		
1000	6T or 5" x 12" - SERIES		
1010	2" x 5-1/3" x 12" Soap Stretcher (Solid Back)	SF	28.00
1020	4" x 5-1/2" x 12" 1 Face Stretcher	"	30.00
1030	2 Face Stretcher	"	35.25
1040	6" x 5-1/3" x 12" 1 Face Stretcher	"	36.00
1050	8" x 5-1/3" x 12" 1 Face Stretcher	"	20.60
2000	8W or 8" x 16" - SERIES		
2010	2" x 8" x 16" Soap Stretcher (Solid Back)	SF	21.33
2020	4" x 8" x 16" 1 Face Stretcher	"	21.84
2030	2 Face Stretcher	"	28.75
2040	6" x 8" x 16" 1 Face Stretcher	"	25.65
2050	8" x 8" x 16" 1 Face Stretcher	"	27.00
2060	Add for Shapes (Average)	PCT	57.75
2070	Add for Designer Colors	"	23.10
2080	Add for Less than Truckload Lots	"	11.55
2090	Add for Base Only	"	28.88
04999.50	**GLASS UNITS**		
0010	4" x 8" x 4"	SF	70.75
0020	6" x 6" x 4"	"	75.50
0030	8" x 8" x 4"	"	45.75
0040	12" x 12" x 4"	"	46.25
04999.60	**TERRA-COTTA**		
0010	Unglazed	SF	15.88
0020	Glazed	"	21.40
0030	Colored Glazed	"	25.60
04999.70	**FLUE LINING**		
0010	8" x 12"	SF	26.50
0020	12" x 12"	"	28.50
0030	16" x 16"	"	45.75
0040	18" x 18"	"	46.50
0050	20" x 20"	"	92.25
0060	24" x 24"	"	117
04999.80	**NATURAL STONE**		
1000	CUT STONE BY S.F.		
1010	Limestone - Indiana and Alabama - 3"	SF	53.75
1020	4"	"	59.50
1030	Minnesota, Wisc, Texas, etc.- 3"	"	60.25
1040	4"	"	68.75
1200	Marble - 2"	"	74.50
1210	3"	"	79.00
1300	Granite - 2"	"	69.50
1310	3"	"	74.75
1400	Slate - 1-1/2"	"	74.25
2000	Ashlar - 4" Sawed Bed		
2100	Limestone - Indiana - Random	SF	42.00
2110	Coursed - 2" - 5" - 8"	"	40.50
2120	Minnesota, Alabama, Wisconsin, etc.		
2130	Split Face - Random	SF	42.00
2140	Coursed	"	47.50
2150	Sawed or Planed Face - Random	"	45.50
2160	Coursed	"	48.00

		UNIT	COST
04999.80	**NATURAL STONE, Cont'd...**		
2200	Marble - Sawed - Random	SF	59.75
2210	Coursed	"	60.75
2300	Granite - Bushhammered - Random	"	61.00
2310	Coursed	"	67.75
2400	Quartzite	"	46.75
3000	ROUGH STONE		
3100	Rubble and Flagstone	SF	43.25
3200	Field Stone or Boulders	"	40.50
3300	Light Weight Boulders (Igneous)		
3310	2" to 4" Veneer - Sawed Back	SF	34.50
3320	3" to 10" Boulders	"	37.00
3330	CUT STONE BY C.F		
3400	Limestone - Indiana and Alabama - 3"	CF	233
3410	4"	"	170
3420	Minnesota, Wisc, Texas, etc.- 3"	"	240
3430	4"	"	230
3500	Marble - 2"	"	520
3510	3"	"	320
3600	Granite - 2"	"	480
3610	3"	"	320
3700	Slate - 1 1/2"	"	490
3800	ASHLAR Limestone - Indiana - Random	"	124
3810	Coursed - 2" - 5" - 8"	"	124
3820	Split Face - Random	"	126
3830	Coursed	"	155
3840	Sawed or Planed Face - Random	"	154
3850	Coursed	"	157
3900	Marble - Sawed - Random	"	187
3910	Coursed	"	189
4000	Granite - Bushhammered - Random	"	190
4010	Coursed	"	195
4100	Quartzite	"	154
4200	ROUGH STONE, Rubble and Flagstone	"	126
4300	Field Stone or Boulders	"	124
04999.90	**PRECAST VENEERS AND SIMULATED MASONRY**		
1000	ARCHITECTURAL PRECAST STONE		
1005	Limestone	SF	50.00
1010	Marble	"	59.75
2010	PRECAST CONCRETE	"	40.75
3010	MOSAIC GRANITE PANELS	"	56.50
04999.91	**MORTAR**		
0010	Portland Cement and Lime Mortar - Labor in Unit Costs	SF	152
0020	Masonry Cement Mortar - Labor in Unit Costs	"	146
0030	Mortar for Standard Brick - Material Only	"	0.63
04999.92	**CORE FILLING FOR REINFORCED CONCRETE**		
1010	Add to Block Prices in 0402.0 (Job Mixed)		
1020	12" x 8" x 16" Plain (includes #4 Rod 16" O.C. Vertical)	SF	4.84
1030	Bond Beam (includes 2 #4 Rods)	"	5.70
1040	8" x 8" x 6" Plain (includes #4 Rod 16" O.C. Vertical)	"	2.96
1050	Bond Beam (includes 2 #4 Rods)	"	2.40
1060	6" x 8" x 16" Plain	"	2.61

		UNIT	COST
04999.92	**CORE FILLING FOR REINFORCED CONCRETE BLOCK, Cont'd...**		
1070	Bond Beam (includes 1 #4 Rod)	SF	1.34
1080	10" x 8" x 16" Bond Beam (includes 2 #4 Rods)	"	5.17
1090	14" x 8" x 16" Bond Beam (includes 2 #4 Rods)	"	5.91
04999.93	**CORE AND CAVITY FILL FOR INSULATED**		
2100	LOOSE		
2110	Core Fill - Expanded Styrene		
2120	12" x 8" x 16" Concrete Block	SF	1.80
2130	10" x 8" x 16" Concrete Block	"	1.60
2140	8" x 8" x 16" Concrete Block	"	1.14
2150	6" x 8" x 16" Concrete Block	"	0.98
2160	8" Brick - Jumbo - Thru the Wall	"	0.86
2170	6" Brick - Jumbo - Thru the Wall	"	0.70
2180	4" Brick - Jumbo	"	0.59
2190	Cavity Fill - per Inch		
2200	Expanded Styrene	SF	0.76
2210	Mica	"	1.28
2220	Fiber Glass	"	0.79
2230	Rock Wool	"	0.79
2240	Cellulose	"	0.77
2300	RIGID - Fiber Glass - 3# Density, 1"	"	1.33
2310	2"	"	1.58
2320	Expanded Styrene - Molded - 1# Density, 1"	SF	0.97
2330	2"	"	1.45
2340	Extruded - 2# Density, 1"	"	1.40
2350	2"	"	2.13
2360	Expanded Urethane - 1"	"	1.51
2370	2"	"	2.00
2380	Perlite - 1"	"	1.45
2390	2"	"	2.26
2400	Add for Embedded Water Type	"	2.26
2410	Add for Glued Applications	"	0.08
04999.95	**CLEANING AND POINTING**		
4010	Brick and Stone - Acid or Chemicals	SF	1.05
4020	Soap and Water	"	1.00
4030	Block & Facing Tile - 1 Face (Incl. Hollow Metal Frames)	"	0.54
4040	Point with White Cement	"	2.10
04999.97	**SCAFFOLD**		
7100	EQUIPMENT, TOOLS AND BLADES		
7110	Percentage of Labor as an average	PCT	7.00
7120	See Division 1 for Rental Rates & New Costs		
7200	SCAFFOLD- Tubular Frame - to 40 feet - Exterior	SF	1.31
7210	Tubular Frame - to 16 feet - Interior	"	1.26
7220	Swing Stage - 40 feet and up	"	1.40

TABLE OF CONTENTS   PAGE

		UNIT	LABOR	MAT.	TOTAL
05050.10	**STRUCTURAL WELDING**				
0080	Welding				
0100	Single pass				
0120	1/8"	LF	3.60	0.33	3.93
0140	3/16"	"	4.80	0.55	5.35
0160	1/4"	"	6.00	0.77	6.77
05050.30	**MECHANICAL ANCHORS**				
0010	Drop-in, stainless steel, for masonry, 3/8"	EA			1.51
0020	1/2"	"			1.76
0030	Hollow wall, for use in gypsum drywall, 6-32"x				
0040	1-1/4"	EA			0.36
0050	1-1/2"	"			0.41
0060	1-3/4"	"			0.38
0070	2"	"			0.36
0080	2-3/8"	"			0.38
1000	Concrete anchor, 3/16"x				
1010	1-1/4"	EA			0.47
1020	1-3/4"	"			0.59
1030	2-3/4"	"			0.59
1040	1/4"x				
1050	1-1/4"	EA			0.59
1060	2-1/4"	"			0.71
1070	2-3/4"	"			0.78
1080	Toggle anchor, 5/8"	"			1.62
2000	3/4"	"			1.62
2010	Toggle bolts, 1/8"x				
2020	2"	EA			0.29
2030	3"	"			0.29
2040	4"	"			0.29
2050	3/16"x				
2060	2"	EA			0.59
2070	3"	"			0.59
2080	4"	"			0.59
2090	1/4"x				
3000	2"	EA			0.86
3010	3"	"			0.86
3020	4"	"			0.86
3030	6"	"			1.69
3050	Sleeve anchor, 3/8"x				
3060	1-7/8"	EA			0.48
3070	2-1/4"	"			0.55
3080	3"	"			0.63
3090	1/4x2-1/4"	"			0.52
4000	1/2"x				
4010	2-1/4"	EA			0.63
4020	3"	"			0.70
4030	4"	"			0.93
4040	5/8"x				
4050	2-1/4"	EA			1.11
4060	4-1/4"	"			1.11
4070	6"	"			1.19
4080	Machine screw, corrosion resistant, for use in masonry, 5/8"x				

05050.30	MECHANICAL ANCHORS, Cont'd...	UNIT	LABOR	MAT.	TOTAL
4090	2"	EA			1.29
5000	4"	"			1.93
5010	6"	"			3.24
5020	8"	"			4.67
5030	Wedge anchor, 1/4"x2-1/4"	"			0.55
5040	3/8"x				
5050	2-1/4"	EA			0.55
5060	3"	"			0.71
5070	4"	"			0.77
5080	5"	"			0.95
5090	1/2"x				
6000	2-1/2"	EA			0.75
6010	3-1/4"	"			1.40
6020	4-1/4"	"			1.66
6030	5-1/2"	"			2.06
6040	5/8"x				
6050	6"	EA			2.91
6060	7"	"			3.47
6070	8"	"			3.99
6080	Spring wing, toggle bolt, for use in hollow walls, 1/8"x				
6090	2"	EA			0.42
7000	3"	"			0.42
7010	4"	"			0.42
7020	3/16"x				
7030	2"	EA			0.59
7040	3"	"			0.59
7050	4"	"			0.59
7060	5"	"			0.59
7070	1/4"x				
7080	2"	EA			0.84
7090	3"	"			0.84
8000	4"	"			0.84
8010	3/8"x6"	"			1.74
8020	Stud anchor, 1-1/4"	"			0.59
8030	3/4"	"			0.59
8040	Hex bolt, zinc plated, 3/4"x				
8050	4"	EA			2.91
8060	6"	"			3.57
8070	8"	"			4.66
8080	1/2"x				
8090	4"	EA			1.02
9000	6"	"			1.54
9010	8"	"			1.98
9020	Lag screw, 1/4"x				
9030	2"	EA			0.24
9040	4"	"			0.50
9050	6"	"			0.78
9060	Machine screw, zinc plated, 5/8"x				
9070	4"	EA			1.45
9100	6"	"			2.45
9110	8"	"			3.55
9120	Ribbed plastic anchor 5/8"x				

		UNIT	LABOR	MAT.	TOTAL
05050.30	**MECHANICAL ANCHORS, Cont'd...**				
9130	1-1/4"	EA			0.30
9140	1-1/2"	"			0.45
05050.90	**METAL ANCHORS**				
1000	Anchor bolts, material only				
1020	3/8" x				
1040	8" long	EA			1.11
1080	12" long	"			1.31
1090	1/2" x				
1100	8" long	EA			1.65
1140	12" long	"			1.93
1170	5/8" x				
1180	8" long	EA			1.54
1220	12" long	"			1.81
1270	3/4" x				
1280	8" long	EA			2.20
1300	12" long	"			2.48
4480	Non-drilling anchor				
4500	1/4"	EA			0.71
4540	3/8"	"			0.88
4560	1/2"	"			1.35
7000	Self-drilling anchor				
7020	1/4"	EA			1.79
7060	3/8"	"			2.68
7080	1/2"	"			3.58
05050.95	**METAL LINTELS**				
0080	Lintels, steel				
0100	Plain	LB	1.80	1.33	3.13
0120	Galvanized	"	1.80	2.00	3.80
05120.10	**BEAMS, GIRDERS, COLUMNS, TRUSSES**				
0100	Beams and girders, A-36				
0120	Welded	TON	850	3,160	4,010
0140	Bolted	"	770	3,070	3,840
0180	Columns				
0185	Pipe				
0190	6" dia.	LB	0.85	1.61	2.46
1300	Structural tube				
1310	6" square				
1320	Light sections	TON	1,700	3,770	5,470
05410.10	**METAL FRAMING**				
0100	Furring channel, galvanized				
0110	Beams and columns, 3/4"				
0120	12" o.c.	SF	7.20	0.44	7.64
0140	16" o.c.	"	6.54	0.34	6.88
0150	Walls, 3/4"				
0160	12" o.c.	SF	3.60	0.44	4.04
0170	16" o.c.	"	3.00	0.34	3.34
0172	24" o.c.	"	2.40	0.24	2.64
0173	1-1/2"				
0174	12" o.c.	SF	3.60	0.72	4.32
0175	16" o.c.	"	3.00	0.55	3.55
0176	24" o.c.	"	2.40	0.37	2.77

05410.10	METAL FRAMING, Cont'd...	UNIT	LABOR	MAT.	TOTAL
0177	Stud, load bearing				
0178	16" o.c.				
0179	16 ga.				
0180	2-1/2"	SF	3.20	1.33	4.53
0190	3-5/8"	"	3.20	1.57	4.77
0200	4"	"	3.20	1.63	4.83
0220	6"	"	3.60	2.05	5.65
0280	18 ga.				
0300	2-1/2"	SF	3.20	1.08	4.28
0310	3-5/8"	"	3.20	1.33	4.53
0320	4"	"	3.20	1.39	4.59
0330	6"	"	3.60	1.76	5.36
0350	8"	"	3.60	2.12	5.72
0360	20 ga.				
0370	2-1/2"	SF	3.20	0.60	3.80
0390	3-5/8"	"	3.20	0.72	3.92
0400	4"	"	3.20	0.79	3.99
0420	6"	"	3.60	0.96	4.56
0440	8"	"	3.60	1.15	4.75
0480	24" o.c.				
0490	16 ga.				
0500	2-1/2"	SF	2.76	0.91	3.67
0510	3-5/8"	"	2.76	1.08	3.84
0520	4"	"	2.76	1.15	3.91
0530	6"	"	3.00	1.39	4.39
0540	8"	"	3.00	1.76	4.76
0545	18 ga.				
0550	2-1/2"	SF	2.76	0.72	3.48
0560	3-5/8"	"	2.76	0.84	3.60
0570	4"	"	2.76	0.91	3.67
0580	6"	"	3.00	1.15	4.15
0590	8"	"	3.00	1.39	4.39
0595	20 ga.				
0600	2-1/2"	SF	2.76	0.44	3.20
0610	3-5/8"	"	2.76	0.49	3.25
0620	4"	"	2.76	0.55	3.31
0630	6"	"	3.00	0.71	3.71
0640	8"	"	3.00	0.88	3.88
05510.10	**STAIRS**				
1000	Stock unit, steel, complete, per riser				
1010	Tread				
1020	3'-6" wide	EA	90.00	240	330
1040	4' wide	"	100	280	380
1060	5' wide	"	120	330	450
1200	Metal pan stair, cement filled, per riser				
1220	3'-6" wide	EA	72.00	260	332
1240	4' wide	"	80.00	300	380
1260	5' wide	"	90.00	340	430
1280	Landing, steel pan	SF	18.00	100	118
1300	Cast iron tread, steel stringers, stock units, per riser				
1310	Tread				

		UNIT	LABOR	MAT.	TOTAL
05510.10	**STAIRS, Cont'd...**				
1320	3'-6" wide	EA	90.00	460	550
1340	4' wide	"	100	530	630
1360	5' wide	"	120	640	760
1400	Stair treads, abrasive, 12" x 3'-6"				
1410	Cast iron				
1420	3/8"	EA	36.00	220	256
1440	1/2"	"	36.00	280	316
1450	Cast aluminum				
1460	5/16"	EA	36.00	250	286
1480	3/8"	"	36.00	270	306
1500	1/2"	"	36.00	320	356
05515.10	**LADDERS**				
0100	Ladder, 18" wide				
0110	With cage	LF	48.00	110	158
0120	Without cage	"	36.00	70.00	106
05520.10	**RAILINGS**				
0080	Railing, pipe				
0090	1-1/4" diameter, welded steel				
0095	2-rail				
0100	Primed	LF	14.50	31.75	46.25
0120	Galvanized	"	14.50	40.75	55.25
0130	3-rail				
0140	Primed	LF	18.00	40.75	58.75
0160	Galvanized	"	18.00	53.00	71.00
0170	Wall mounted, single rail, welded steel				
0180	Primed	LF	11.00	21.25	32.25
0200	Galvanized	"	11.00	27.50	38.50
0210	1-1/2" diameter, welded steel				
0215	2-rail				
0220	Primed	LF	14.50	34.50	49.00
0240	Galvanized	"	14.50	44.75	59.25
0245	3-rail				
0250	Primed	LF	18.00	43.25	61.25
0260	Galvanized	"	18.00	56.00	74.00
0270	Wall mounted, single rail, welded steel				
0280	Primed	LF	11.00	21.75	32.75
0300	Galvanized	"	11.00	28.50	39.50
0960	2" diameter, welded steel				
0980	2-rail				
1000	Primed	LF	16.00	41.25	57.25
1020	Galvanized	"	16.00	54.00	70.00
1030	3-rail				
1040	Primed	LF	20.50	52.00	72.50
1070	Galvanized	"	20.50	68.00	88.50
1075	Wall mounted, single rail, welded steel				
1080	Primed	LF	12.00	23.75	35.75
1100	Galvanized	"	12.00	30.75	42.75
05580.10	**METAL SPECIALTIES**				
0060	Kick plate				
0080	4" high x 1/4" thick				
0100	Primed	LF	14.50	8.47	22.97

		UNIT	LABOR	MAT.	TOTAL
05580.10	**METAL SPECIALTIES, Cont'd...**				
0120	Galvanized	LF	14.50	9.62	24.12
0130	6" high x 1/4" thick				
0140	Primed	LF	16.00	9.58	25.58
0160	Galvanized	"	16.00	11.50	27.50
0200	Fire Escape (10'-12' high)				
0210	Landing with fixed stair, 3'-6" wide	EA	1,440	5,580	7,020
0220	4'-6" wide	"	1,440	6,310	7,750
05700.10	**ORNAMENTAL METAL**				
1020	Railings, square bars, 6" o.c., shaped top rails				
1040	Steel	LF	36.00	92.00	128
1060	Aluminum	"	36.00	110	146
1080	Bronze	"	48.00	230	278
1100	Stainless steel	"	48.00	240	288
1200	Laminated metal or wood handrails				
1220	2-1/2" round or oval shape	LF	36.00	280	316

		UNIT	COST
05999.10	**STRUCTURAL STEEL FRAME**		
1000	To 30 Ton 20' Span	SF	13.26
1010	24' Span	"	14.01
1020	28' Span	"	14.06
1030	32' Span	"	17.00
1040	36' Span	"	19.00
2010	44' Span	"	21.62
2020	48' Span	"	23.31
2030	52' Span	"	25.72
2050	60' Span	"	30.00
3000	Deduct for Over 30 Ton	PCT	0.05
05999.40	**OPEN WEB JOISTS**		
1000	To 20 Ton, 20' Span	SF	5.12
1010	24' Span	"	5.15
1030	32' Span	"	5.25
2000	40' Span	"	6.02
2010	48' Span	"	7.09
2030	56' Span	"	8.25
2040	60' Span	"	8.84
3000	Deduct for Over 20 Ton	PCT	0.10
05999.50	**METAL DECKING**		
1000	1/2" Deep - Ribbed - Baked Enamel - 18 Ga	SF	5.82
1020	22 Ga	"	4.90
2000	3" Deep - Ribbed - Baked Enamel - 18 Ga	"	11.59
2020	22 Ga	"	10.20
3000	4-1/2" Deep - Ribbed - Baked Enamel - 16 Ga	"	15.22
3010	18 Ga	"	13.25
3020	20 Ga	"	12.26
4000	3" Deep - Cellular - 18 Ga	"	13.89
4010	16 Ga	"	17.07
4020	4-1/2" Deep - Cellular - 18 Ga	"	19.50
4030	16 Ga	"	22.48
4040	Add for Galvanized	PCT	15.00
4050	Corrugated Black Standard .015	SF	2.68
4060	Heavy Duty - 26 Ga	"	2.89
4070	S. Duty - 24 Ga	"	3.79
4080	22 Ga	"	3.94
5000	Add for Galvanized	PCT	15.00
05999.60	**METAL SIDING AND ROOFING**		
1000	Aluminum- Anodized	SF	7.32
1010	Porcelainized	"	14.58
1020	Corrugated	"	4.97
1030	Enamel, Baked Ribbed	"	7.78
1040	24 Ga Ribbed	"	5.32
1050	Acrylic	"	9.08
2000	Porcelain Ribbed	"	10.03
2010	Galvanized - Corrugated	"	5.01
2020	Plastic Faced	"	10.91
2030	Protected Metal	"	13.28
2040	Add for Liner Panels	"	4.12
2050	Add for Insulation	"	0.64

TABLE OF CONTENTS

		UNIT	LABOR	MAT.	TOTAL
06110.10	**BLOCKING**				
1215	Wood construction				
1220	Walls				
1230	2x4	LF	3.62	0.60	4.22
1240	2x6	"	4.07	0.92	4.99
1250	2x8	"	4.34	1.21	5.55
1260	2x10	"	4.66	1.62	6.28
1270	2x12	"	5.01	2.09	7.10
1280	Ceilings				
1290	2x4	LF	4.07	0.60	4.67
1300	2x6	"	4.66	0.92	5.58
1310	2x8	"	5.01	1.21	6.22
1320	2x10	"	5.43	1.62	7.05
1330	2x12	"	5.93	2.09	8.02
06110.20	**CEILING FRAMING**				
1000	Ceiling joists				
1070	16" o.c.				
1080	2x4	SF	1.25	0.73	1.98
1090	2x6	"	1.30	1.09	2.39
1100	2x8	"	1.35	1.55	2.90
1110	2x10	"	1.41	1.74	3.15
1120	2x12	"	1.48	3.26	4.74
1130	24" o.c.				
1140	2x4	SF	1.03	0.52	1.55
1150	2x6	"	1.08	0.87	1.95
1160	2x8	"	1.14	1.30	2.44
1170	2x10	"	1.20	1.55	2.75
1180	2x12	"	1.27	3.92	5.19
1200	Headers and nailers				
1210	2x4	LF	2.10	0.60	2.70
1220	2x6	"	2.17	0.92	3.09
1230	2x8	"	2.33	1.21	3.54
1240	2x10	"	2.50	1.62	4.12
1250	2x12	"	2.71	1.99	4.70
1300	Sister joists for ceilings				
1310	2x4	LF	4.66	0.60	5.26
1320	2x6	"	5.43	0.92	6.35
1330	2x8	"	6.52	1.21	7.73
1340	2x10	"	8.15	1.62	9.77
1350	2x12	"	10.75	1.99	12.74
06110.30	**FLOOR FRAMING**				
1000	Floor joists				
1180	16" o.c.				
1190	2x6	SF	1.08	0.95	2.03
1200	2x8	"	1.10	1.34	2.44
1220	2x10	"	1.12	1.63	2.75
1230	2x12	"	1.16	2.03	3.19
1240	2x14	"	1.20	4.52	5.72
1250	3x6	"	1.12	3.13	4.25
1260	3x8	"	1.16	4.00	5.16
1270	3x10	"	1.20	5.04	6.24
1280	3x12	"	1.25	6.09	7.34

06110.30 FLOOR FRAMING, Cont'd...

		UNIT	LABOR	MAT.	TOTAL
1290	3x14	SF	1.30	7.21	8.51
1300	4x6	"	1.12	4.00	5.12
1310	4x8	"	1.16	5.48	6.64
1320	4x10	"	1.20	6.78	7.98
1330	4x12	"	1.25	8.00	9.25
1340	4x14	"	1.30	9.57	10.87
2000	Sister joists for floors				
2010	2x4	LF	4.07	0.60	4.67
2020	2x6	"	4.66	0.92	5.58
2030	2x8	"	5.43	1.21	6.64
2040	2x10	"	6.52	1.62	8.14
2050	2x12	"	8.15	2.09	10.24
2060	3x6	"	6.52	3.04	9.56
2070	3x8	"	7.24	3.73	10.97
2080	3x10	"	8.15	4.95	13.10
2090	3x12	"	9.32	5.99	15.31
2100	4x6	"	6.52	3.92	10.44
2110	4x8	"	7.24	5.22	12.46
2120	4x10	"	8.15	6.78	14.93
2130	4x12	"	9.32	7.56	16.88

06110.40 FURRING

		UNIT	LABOR	MAT.	TOTAL
1100	Furring, wood strips				
1102	Walls				
1105	On masonry or concrete walls				
1107	1x2 furring				
1110	12" o.c.	SF	2.03	0.48	2.51
1120	16" o.c.	"	1.86	0.41	2.27
1130	24" o.c.	"	1.71	0.40	2.11
1135	1x3 furring				
1140	12" o.c.	SF	2.03	0.60	2.63
1150	16" o.c.	"	1.86	0.55	2.41
1160	24" o.c.	"	1.71	0.42	2.13
1165	On wood walls				
1167	1x2 furring				
1170	12" o.c.	SF	1.44	0.48	1.92
1180	16" o.c.	"	1.30	0.41	1.71
1190	24" o.c.	"	1.18	0.38	1.56
1195	1x3 furring				
1200	12" o.c.	SF	1.44	0.62	2.06
1210	16" o.c.	"	1.30	0.52	1.82
1220	24" o.c.	"	1.18	0.42	1.60

06110.50 ROOF FRAMING

		UNIT	LABOR	MAT.	TOTAL
1000	Roof framing				
1005	Rafters, gable end				
1008	0-2 pitch (flat to 2-in-12)				
1070	16" o.c.				
1080	2x6	SF	1.16	1.09	2.25
1090	2x8	"	1.20	1.53	2.73
1100	2x10	"	1.25	1.74	2.99
1110	2x12	"	1.30	3.21	4.51
1120	24" o.c.				

		UNIT	LABOR	MAT.	TOTAL
06110.50	**ROOF FRAMING, Cont'd...**				
1130	2x6	SF	0.98	0.60	1.58
1140	2x8	"	1.01	1.27	2.28
1150	2x10	"	1.05	1.48	2.53
1160	2x12	"	1.08	2.60	3.68
1165	4-6 pitch (4-in-12 to 6-in-12)				
1220	16" o.c.				
1230	2x6	SF	1.20	1.09	2.29
1240	2x8	"	1.25	1.74	2.99
1250	2x10	"	1.30	1.99	3.29
1260	2x12	"	1.35	2.96	4.31
1270	24" o.c.				
1280	2x6	SF	1.01	0.87	1.88
1290	2x8	"	1.05	1.48	2.53
1300	2x10	"	1.12	1.56	2.68
1310	2x12	"	1.25	2.43	3.68
1315	8-12 pitch (8-in-12 to 12-in-12)				
1380	16" o.c.				
1390	2x6	SF	1.25	1.21	2.46
1400	2x8	"	1.30	1.95	3.25
1410	2x10	"	1.35	2.17	3.52
1420	2x12	"	1.41	3.13	4.54
1430	24" o.c.				
1440	2x6	SF	1.05	0.95	2.00
1450	2x8	"	1.08	1.55	2.63
1460	2x10	"	1.12	1.74	2.86
1470	2x12	"	1.16	2.78	3.94
2000	Ridge boards				
2010	2x6	LF	3.26	0.92	4.18
2020	2x8	"	3.62	1.21	4.83
2030	2x10	"	4.07	1.62	5.69
2040	2x12	"	4.66	2.09	6.75
3000	Hip rafters				
3010	2x6	LF	2.33	0.92	3.25
3020	2x8	"	2.41	1.21	3.62
3030	2x10	"	2.50	1.62	4.12
3040	2x12	"	2.60	2.09	4.69
3180	Jack rafters				
3190	4-6 pitch (4-in-12 to 6-in-12)				
3200	16" o.c.				
3210	2x6	SF	1.91	1.13	3.04
3220	2x8	"	1.97	1.74	3.71
3230	2x10	"	2.10	1.99	4.09
3240	2x12	"	2.17	2.96	5.13
3250	24" o.c.				
3260	2x6	SF	1.48	0.87	2.35
3270	2x8	"	1.51	1.48	2.99
3280	2x10	"	1.59	1.74	3.33
3290	2x12	"	1.63	2.52	4.15
3295	8-12 pitch (8-in-12 to 12-in-12)				
3300	16" o.c.				
3310	2x6	SF	2.03	1.74	3.77
3320	2x8	"	2.10	2.17	4.27

		UNIT	LABOR	MAT.	TOTAL
06110.50	**ROOF FRAMING, Cont'd...**				
3330	2x10	SF	2.17	3.13	5.30
3340	2x12	"	2.24	4.34	6.58
3350	24" o.c.				
3360	2x6	SF	1.55	1.38	2.93
3370	2x8	"	1.59	1.74	3.33
3380	2x10	"	1.63	2.78	4.41
3390	2x12	"	1.67	4.00	5.67
4980	Sister rafters				
5000	2x4	LF	4.66	0.60	5.26
5010	2x6	"	5.43	0.92	6.35
5020	2x8	"	6.52	1.21	7.73
5030	2x10	"	8.15	1.62	9.77
5040	2x12	"	10.75	2.09	12.84
5050	Fascia boards				
5060	2x4	LF	3.26	0.60	3.86
5070	2x6	"	3.26	0.92	4.18
5080	2x8	"	3.62	1.21	4.83
5090	2x10	"	3.62	1.62	5.24
5100	2x12	"	4.07	2.09	6.16
7980	Cant strips				
7985	Fiber				
8000	3x3	LF	1.86	0.48	2.34
8020	4x4	"	1.97	0.67	2.64
8030	Wood				
8040	3x3	LF	1.97	2.52	4.49
06110.60	**SLEEPERS**				
0960	Sleepers, over concrete				
1090	16" o.c.				
1100	1x2	SF	1.30	0.26	1.56
1120	1x3	"	1.30	0.37	1.67
1140	2x4	"	1.55	0.79	2.34
1160	2x6	"	1.63	1.17	2.80
06110.65	**SOFFITS**				
0980	Soffit framing				
1000	2x3	LF	4.66	0.41	5.07
1020	2x4	"	5.01	0.51	5.52
1030	2x6	"	5.43	0.75	6.18
1040	2x8	"	5.93	1.06	6.99
06110.70	**WALL FRAMING**				
0960	Framing wall, studs				
1110	16" o.c.				
1120	2x3	SF	1.01	0.43	1.44
1140	2x4	"	1.01	0.61	1.62
1150	2x6	"	1.08	0.87	1.95
1160	2x8	"	1.12	1.37	2.49
1165	24" o.c.				
1170	2x3	SF	0.88	0.34	1.22
1180	2x4	"	0.88	0.46	1.34
1190	2x6	"	0.93	0.73	1.66
1200	2x8	"	0.95	0.95	1.90
1480	Plates, top or bottom				

DIVISION # 06 WOOD AND PLASTICS

		UNIT	LABOR	MAT.	TOTAL
06110.70	**WALL FRAMING, Cont'd...**				
1500	2x3	LF	1.91	0.41	2.32
1510	2x4	"	2.03	0.51	2.54
1520	2x6	"	2.17	0.75	2.92
1530	2x8	"	2.33	1.06	3.39
2000	Headers, door or window				
2044	2x8				
2046	Single				
2050	4' long	EA	40.75	4.51	45.26
2060	8' long	"	50.00	9.01	59.01
2065	Double				
2070	4' long	EA	46.50	9.01	55.51
2080	8' long	"	59.00	18.00	77.00
2134	2x12				
2138	Single				
2140	6' long	EA	50.00	9.89	59.89
2150	12' long	"	65.00	19.50	84.50
2155	Double				
2160	6' long	EA	59.00	19.50	78.50
2170	12' long	"	72.00	38.75	111
06115.10	**FLOOR SHEATHING**				
1980	Sub-flooring, plywood, CDX				
2000	1/2" thick	SF	0.81	0.61	1.42
2020	5/8" thick	"	0.93	0.88	1.81
2080	3/4" thick	"	1.08	1.62	2.70
2090	Structural plywood				
2100	1/2" thick	SF	0.81	0.96	1.77
2120	5/8" thick	"	0.93	1.54	2.47
2140	3/4" thick	"	1.00	1.62	2.62
5990	Underlayment				
6000	Hardboard, 1/4" tempered	SF	0.81	0.90	1.71
6010	Plywood, CDX				
6020	3/8" thick	SF	0.81	0.94	1.75
6040	1/2" thick	"	0.86	1.12	1.98
6060	5/8" thick	"	0.93	1.30	2.23
6080	3/4" thick	"	1.00	1.62	2.62
06115.20	**ROOF SHEATHING**				
0080	Sheathing				
0090	Plywood, CDX				
1000	3/8" thick	SF	0.84	0.94	1.78
1020	1/2" thick	"	0.86	1.12	1.98
1040	5/8" thick	"	0.93	1.30	2.23
1060	3/4" thick	"	1.00	1.62	2.62
1080	Structural plywood				
2040	3/8" thick	SF	0.84	0.59	1.43
2060	1/2" thick	"	0.86	0.77	1.63
2080	5/8" thick	"	0.93	0.94	1.87
2100	3/4" thick	"	1.00	1.13	2.13
06115.30	**WALL SHEATHING**				
0980	Sheathing				
0990	Plywood, CDX				
1000	3/8" thick	SF	0.96	0.94	1.90

		UNIT	LABOR	MAT.	TOTAL
06115.30	**WALL SHEATHING, Cont'd...**				
1020	1/2" thick	SF	1.00	1.12	2.12
1040	5/8" thick	"	1.08	1.30	2.38
1060	3/4" thick	"	1.18	1.62	2.80
3000	Waferboard				
3020	3/8" thick	SF	0.96	0.59	1.55
3040	1/2" thick	"	1.00	0.77	1.77
3060	5/8" thick	"	1.08	0.94	2.02
3080	3/4" thick	"	1.18	1.03	2.21
4100	Structural plywood				
4120	3/8" thick	SF	0.96	0.94	1.90
4140	1/2" thick	"	1.00	1.12	2.12
4160	5/8" thick	"	1.08	1.30	2.38
4180	3/4" thick	"	1.18	1.12	2.30
7000	Gypsum, 1/2" thick	"	1.00	0.59	1.59
8000	Asphalt impregnated fiberboard, 1/2" thick	"	1.00	1.03	2.03
06125.10	**WOOD DECKING**				
0090	Decking, T&G solid				
0095	Cedar				
0100	3" thick	SF	1.63	11.75	13.38
0120	4" thick	"	1.73	14.50	16.23
1030	Fir				
1040	3" thick	SF	1.63	5.10	6.73
1060	4" thick	"	1.73	6.19	7.92
1080	Southern yellow pine				
2000	3" thick	SF	1.86	5.10	6.96
2020	4" thick	"	2.00	5.39	7.39
3120	White pine				
3140	3" thick	SF	1.63	6.19	7.82
3160	4" thick	"	1.73	8.38	10.11
06130.10	**HEAVY TIMBER**				
1000	Mill framed structures				
1010	Beams to 20' long				
1020	Douglas fir				
1040	6x8	LF	8.60	8.36	16.96
1042	6x10	"	8.90	9.87	18.77
1044	6x12	"	9.56	11.75	21.31
1046	6x14	"	9.93	14.25	24.18
1048	6x16	"	10.25	15.50	25.75
1060	8x10	"	8.90	13.00	21.90
1070	8x12	"	9.56	15.50	25.06
1080	8x14	"	9.93	17.75	27.68
1090	8x16	"	10.25	20.25	30.50
1380	Columns to 12' high				
1400	Douglas fir				
1420	6x6	LF	13.00	6.01	19.01
1440	8x8	"	13.00	10.25	23.25
1460	10x10	"	14.25	18.00	32.25
1480	12x12	"	14.25	22.25	36.50
2000	Posts, treated				
2100	4x4	LF	2.60	2.07	4.67
2120	6x6	"	3.26	6.01	9.27

		UNIT	LABOR	MAT.	TOTAL

06190.20   WOOD TRUSSES

0960	Truss, fink, 2x4 members				
0980	3-in-12 slope				
1030	5-in-12 slope				
1040	24' span	EA	76.00	130	206
1050	28' span	"	78.00	140	218
1055	30' span	"	81.00	150	231
1060	32' span	"	81.00	160	241
1070	40' span	"	86.00	210	296
1074	Gable, 2x4 members				
1078	5-in-12 slope				
1080	24' span	EA	76.00	150	226
1100	28' span	"	78.00	180	258
1120	30' span	"	81.00	190	271
1160	36' span	"	83.00	210	293
1180	40' span	"	86.00	230	316
1190	King post type, 2x4 members				
2000	4-in-12 slope				
2040	16' span	EA	70.00	91.00	161
2060	18' span	"	72.00	98.00	170
2080	24' span	"	76.00	110	186
2120	30' span	"	81.00	140	221
2160	38' span	"	83.00	180	263
2180	42' span	"	89.00	220	309

06190.30   LAMINATED BEAMS

0010	Parallel strand beams 3-1/2" wide x				
0020	9-1/2"	LF	3.68	12.25	15.93
0030	11-1/4"	"	3.82	13.00	16.82
0040	11-7/8"	"	3.97	13.75	17.72
0050	14"	"	4.69	17.50	22.19
0060	16"	"	5.16	20.75	25.91
0070	18"	"	5.73	24.50	30.23
1000	Laminated veneer beams, 1-3/4" wide x				
1010	11-7/8"	LF	3.97	8.36	12.33
1020	14"	"	4.69	10.50	15.19
1030	16"	"	5.16	10.25	15.41
1040	18"	"	5.73	13.00	18.73
2000	Laminated strand beams, 1-3/4" wide x				
2010	9-1/2"	LF	3.68	5.36	9.04
2020	11-7/8"	"	3.97	6.13	10.10
2030	14"	"	4.69	7.21	11.90
2040	16"	"	5.16	8.27	13.43
2050	3-1/2" wide x				
2060	9-1/2"	LF	3.68	9.43	13.11
2070	11-7/8"	"	3.97	12.25	16.22
2080	14"	"	4.69	14.50	19.19
2090	16"	"	5.16	17.25	22.41
3000	Gluelam beam, 3-1/2" wide x				
3010	10"	LF	3.68	14.25	17.93
3020	12"	"	4.30	16.75	21.05
3030	15"	"	4.91	20.00	24.91
3040	5-1/2" wide x				

		UNIT	LABOR	MAT.	TOTAL
06190.30	**LAMINATED BEAMS, Cont'd...**				
3050	10"	LF	3.68	23.00	26.68
3060	16"	"	5.16	36.00	41.16
3070	20"	"	6.07	42.50	48.57
3100	7-1/2" wide x				
3120	10"	LF	3.97	30.50	34.47
3140	12"	"	5.43	48.00	53.43
3160	15"	"	6.45	57.00	63.45
3180	18"	"	6.88	68.00	74.88
3300	9-1/2" wide x				
3320	10"	LF	3.97	30.50	34.47
3340	16"	"	5.43	49.00	54.43
3360	20"	"	6.45	61.00	67.45
3380	24"	"	6.88	77.00	83.88
3400	30"	"	7.94	96.00	104
06200.10	**FINISH CARPENTRY**				
0070	Mouldings and trim				
0980	Apron, flat				
1000	9/16 x 2	LF	3.26	1.99	5.25
1010	9/16 x 3-1/2	"	3.43	4.59	8.02
1015	Base				
1020	Colonial				
1022	7/16 x 2-1/4	LF	3.26	2.37	5.63
1024	7/16 x 3	"	3.26	3.07	6.33
1026	7/16 x 3-1/4	"	3.26	3.14	6.40
1028	9/16 x 3	"	3.43	3.07	6.50
1030	9/16 x 3-1/4	"	3.43	3.21	6.64
1034	11/16 x 2-1/4	"	3.62	3.37	6.99
1035	Ranch				
1036	7/16 x 2-1/4	LF	3.26	2.60	5.86
1038	7/16 x 3-1/4	"	3.26	3.07	6.33
1039	9/16 x 2-1/4	"	3.43	2.83	6.26
1041	9/16 x 3	"	3.43	3.07	6.50
1043	9/16 x 3-1/4	"	3.43	3.14	6.57
1050	Casing				
1060	11/16 x 2-1/2	LF	2.96	2.44	5.40
1070	11/16 x 3-1/2	"	3.10	2.76	5.86
1180	Chair rail				
1200	9/16 x 2-1/2	LF	3.26	2.60	5.86
1210	9/16 x 3-1/2	"	3.26	3.60	6.86
1250	Closet pole				
1300	1-1/8" dia.	LF	4.34	1.76	6.10
1310	1-5/8" dia.	"	4.34	2.60	6.94
1340	Cove				
1500	9/16 x 1-3/4	LF	3.26	1.99	5.25
1510	11/16 x 2-3/4	"	3.26	3.07	6.33
1550	Crown				
1600	9/16 x 1-5/8	LF	4.34	2.60	6.94
1620	11/16 x 3-5/8	"	5.43	3.07	8.50
1640	11/16 x 5-1/4	"	6.52	5.14	11.66
1680	Drip cap				
1700	1-1/16 x 1-5/8	LF	3.26	2.76	6.02

DIVISION # 06 WOOD AND PLASTICS

06200.10	FINISH CARPENTRY, Cont'd...	UNIT	LABOR	MAT.	TOTAL
1780	Glass bead				
1800	3/8 x 3/8	LF	4.07	0.99	5.06
1840	5/8 x 5/8	"	4.07	1.30	5.37
1860	3/4 x 3/4	"	4.07	1.53	5.60
1880	Half round				
1900	1/2	LF	2.60	1.15	3.75
1910	5/8	"	2.60	1.53	4.13
1920	3/4	"	2.60	2.07	4.67
1980	Lattice				
2000	1/4 x 7/8	LF	2.60	0.92	3.52
2010	1/4 x 1-1/8	"	2.60	0.99	3.59
2030	1/4 x 1-3/4	"	2.60	1.19	3.79
2040	1/4 x 2	"	2.60	1.38	3.98
2080	Ogee molding				
2100	5/8 x 3/4	LF	3.26	1.83	5.09
2110	11/16 x 1-1/8	"	3.26	4.30	7.56
2120	11/16 x 1-3/8	"	3.26	3.37	6.63
2180	Parting bead				
2200	3/8 x 7/8	LF	4.07	1.53	5.60
2300	Quarter round				
2301	1/4 x 1/4	LF	2.60	0.54	3.14
2303	3/8 x 3/8	"	2.60	0.76	3.36
2305	1/2 x 1/2	"	2.60	0.99	3.59
2307	11/16 x 11/16	"	2.83	0.99	3.82
2309	3/4 x 3/4	"	2.83	1.83	4.66
2311	1-1/16 x 1-1/16	"	2.96	1.45	4.41
2380	Railings, balusters				
2400	1-1/8 x 1-1/8	LF	6.52	4.91	11.43
2410	1-1/2 x 1-1/2	"	5.93	5.75	11.68
2480	Screen moldings				
2500	1/4 x 3/4	LF	5.43	1.22	6.65
2510	5/8 x 5/16	"	5.43	1.53	6.96
2580	Shoe				
2600	7/16 x 11/16	LF	2.60	1.53	4.13
2605	Sash beads				
2620	1/2 x 7/8	LF	5.43	1.99	7.42
2640	5/8 x 7/8	"	5.93	2.15	8.08
2760	Stop				
2780	5/8 x 1-5/8				
2800	Colonial	LF	4.07	1.06	5.13
2810	Ranch	"	4.07	1.06	5.13
2880	Stools				
2900	11/16 x 2-1/4	LF	7.24	4.68	11.92
2910	11/16 x 2-1/2	"	7.24	4.91	12.15
2920	11/16 x 5-1/4	"	8.15	5.06	13.21
4000	Exterior trim, casing, select pine, 1x3	"	3.26	3.37	6.63
4010	Douglas fir				
4020	1x3	LF	3.26	1.60	4.86
4040	1x4	"	3.26	1.99	5.25
4060	1x6	"	3.62	2.60	6.22
4100	1x8	"	4.07	3.60	7.67
5000	Cornices, white pine, #2 or better				

DIVISION # 06 WOOD AND PLASTICS

		UNIT	LABOR	MAT.	TOTAL
06200.10	**FINISH CARPENTRY, Cont'd...**				
5040	1x4	LF	3.26	1.22	4.48
5080	1x8	"	3.83	2.44	6.27
5120	1x12	"	4.34	3.91	8.25
8600	Shelving, pine				
8620	1x8	LF	5.01	1.76	6.77
8640	1x10	"	5.21	2.30	7.51
8660	1x12	"	5.43	2.91	8.34
8800	Plywood shelf, 3/4", with edge band, 12" wide	"	6.52	3.14	9.66
8840	Adjustable shelf, and rod, 12" wide				
8860	3' to 4' long	EA	16.25	25.00	41.25
8880	5' to 8' long	"	21.75	47.00	68.75
8900	Prefinished wood shelves with brackets and supports				
8905	8" wide				
8910	3' long	EA	16.25	74.00	90.25
8922	4' long	"	16.25	85.00	101
8924	6' long	"	16.25	120	136
8930	10" wide				
8940	3' long	EA	16.25	81.00	97.25
8942	4' long	"	16.25	120	136
8946	6' long	"	16.25	130	146
06220.10	**MILLWORK**				
0070	Countertop, laminated plastic				
0080	25" x 7/8" thick				
0099	Minimum	LF	16.25	18.75	35.00
0100	Average	"	21.75	35.50	57.25
0110	Maximum	"	26.00	52.00	78.00
0115	25" x 1-1/4" thick				
0120	Minimum	LF	21.75	22.75	44.50
0130	Average	"	26.00	45.50	71.50
0140	Maximum	"	32.50	68.00	101
0160	Add for cutouts	EA	40.75		40.75
0165	Backsplash, 4" high, 7/8" thick	LF	13.00	25.00	38.00
2000	Plywood, sanded, A-C				
2020	1/4" thick	SF	2.17	1.61	3.78
2040	3/8" thick	"	2.33	1.75	4.08
2060	1/2" thick	"	2.50	1.98	4.48
2070	A-D				
2080	1/4" thick	SF	2.17	1.53	3.70
2090	3/8" thick	"	2.33	1.75	4.08
2100	1/2" thick	"	2.50	1.90	4.40
2500	Base cabinet, 34-1/2" high, 24" deep, hardwood				
2540	Minimum	LF	26.00	250	276
2560	Average	"	32.50	280	313
2580	Maximum	"	43.50	310	354
2600	Wall cabinets				
2640	Minimum	LF	21.75	75.00	96.75
2660	Average	"	26.00	100	126
2680	Maximum	"	32.50	130	163
06300.10	**WOOD TREATMENT**				
1000	Creosote preservative treatment				
1020	8 lb/cf	BF			0.74

		UNIT	LABOR	MAT.	TOTAL
06300.10	**WOOD TREATMENT, Cont'd...**				
1040	10 lb/cf	BF			0.89
1060	Salt preservative treatment				
1070	Oil borne				
1080	Minimum	BF			0.68
1100	Maximum	"			0.96
1120	Water borne				
1140	Minimum	BF			0.48
1150	Maximum	"			0.74
1200	Fire retardant treatment				
1220	Minimum	BF			0.96
1240	Maximum	"			1.16
1300	Kiln dried, softwood, add to framing costs				
1320	1" thick	BF			0.34
1360	3" thick	"			0.61
06430.10	**STAIRWORK**				
0080	Risers, 1x8, 42" wide				
0100	White oak	EA	32.50	53.00	85.50
0120	Pine	"	32.50	47.00	79.50
0130	Treads, 1-1/16" x 9-1/2" x 42"				
0140	White oak	EA	40.75	63.00	104
06440.10	**COLUMNS**				
0980	Column, hollow, round wood				
0990	12" diameter				
1000	10' high	EA	95.00	940	1,035
1080	16' high	"	140	1,710	1,850
2000	24" diameter				
2020	16' high	EA	140	3,900	4,040
2060	20' high	"	150	5,460	5,610
2100	24' high	"	160	6,270	6,430

		UNIT	COST
06999.10	**ROUGH CARPENTRY**		
1000	LIGHT FRAMING AND SHEATHING		
1100	Joists and Headers - Floor Area - 16" O.C.		
1110	2" x 6" Joists - with Headers & Bridging	SF	3.67
1120	2" x 8" Joists	"	4.52
1130	2" x 10" Joists	"	5.03
1140	2" x 12" Joists	"	5.99
1150	Add for Ceiling Joists, 2nd Floor and Above	"	0.13
1160	Add for Sloped Installation	"	0.23
1200	Studs, Plates and Framing - 8' Wall Height - 16" O.C.	"	2.20
1210	2" x 3" Stud Wall - Non-Bearing (Single Top Plate)		
1220	2" x 4" Stud Wall - Bearing (Double Top Plate)	SF	2.64
1230	2" x 4" Stud Wall - Non-Bearing (Single Top Plate)	"	1.95
1240	2" x 6" Stud Wall - Bearing (Double Top Plate)	"	3.57
1250	2" x 6" Stud Wall - Non-Bearing (Single Top Plate)	"	2.65
1300	Add for Stud Wall - 12" O.C.	"	0.12
1310	Deduct for Stud Wall - 24" O.C.	"	0.33
1320	Add for Bolted Plates or Sills	"	0.20
1330	Add for Each Foot Above 8'	"	0.11
1340	Add to Above for Fire Stops, Fillers and Nailers	"	0.70
1350	Add for Soffits and Suspended Framing	"	0.95
1360	Bridging - 1" x 3" Wood Diagonal	EA	3.72
1370	2" x 8" Solid	"	4.14
1400	Rafters		
1410	2" x 4" Rafter (Incl Bracing) 3 - 12 Slope	SF	2.71
1420	4 - 12 Slope	"	2.71
1430	5 - 12 Slope	"	3.06
1440	6 - 12 Slope	"	3.60
1500	Add for Hip-and-Valley Type	"	0.71
1510	Add for 1' 0" Overhang - Total Area of Roof	"	0.27
1520	2" x 6" Rafter (Incl Bracing) 3-12 Slope	"	3.43
1530	4-12 Slope	"	3.46
1540	5-12 Slope	"	3.56
1550	6-12 Slope	"	3.76
1560	Add for Hip-and-Valley Type	"	0.80
1600	Stairs	EA	690
1700	Sub Floor Sheathing (Structural)		
1710	1" x 8" and 1" x 10" #3 Pine	SF	2.41
1720	1/2" x 4' x 8' CD Plywood - Exterior	"	1.93
1730	5/8" x 4' x 8'	"	2.35
1740	3/4" x 4' x 8'	"	3.14
1800	Floor Sheathing (Over Sub Floor)		
1810	3/8" x 4' x 8' CD Plywood	SF	1.79
1820	1/2" x 4' x 8'	"	2.16
1830	5/8" x 4' x 8'	"	2.81
1840	1/2" x 4' x 8' Particle Board	"	2.22
1850	5/8" x 4' x 8'	"	2.61
1860	3/4" x 4' x 8'	"	3.20
1900	Wall Sheathing		
1910	1' x 8" and 1" x 10" - #3 Pine	SF	2.46
1920	3/8" x 4' x 8' CD Plywood - Exterior	"	1.84
1930	1/2" x 4' x 8'	"	2.15
1940	5/8" x 4' x 8'	"	2.95

		UNIT	COST
06999.10	**ROUGH CARPENTRY, Cont'd...**		
1950	3/4" x 4' x 8'	SF	3.31
1960	25/32" x 4' x 8' Fiber Board - Impregnated	"	2.06
1970	1" x 2' x 8' T&G Styrofoam	"	1.81
1980	2" x 2' x 8'	"	2.67
2000	Roof Sheathing - Flat Construction		
2010	1" x 6" and 1" x 8" - #3 Pine	SF	3.05
2020	1/2" x 4' x 8' CD Plywood - Exterior	"	2.84
2030	5/8" x 4' x 8'	"	2.48
2040	3/4" x 4' x 8'	"	3.32
2050	Add for Sloped Roof Construction (to 5-12 slope)	"	0.28
2060	Add for Steep Sloped Construction (over 5-12 slope)	"	0.51
2070	Add to Above Sheathing		
2080	AC or AD Plywood	SF	0.51
2090	10' Length Plywood	"	0.39
2100	HEAVY FRAMING		
2200	Columns and Beams - 16' Span Average - Floor Area	SF	10.55
2210	20' Span	"	10.97
2220	24' Span	"	12.95
2300	Deck 2" x 6" T&G - Fir Random Construction Grade	"	6.39
2310	3" x 6" T&G - Fir Random Construction Grade	"	8.52
2320	4" x 6"	"	10.55
2330	2" x 6" T&G - Red Cedar	"	6.74
2340	Add for D Grade Cedar	"	2.64
2350	2" x 6" T&G - Panelized Fir	"	5.69
3000	MISCELLANEOUS CARPENTRY		
3100	Blocking and Bucks (2" x 4" and 2" x 6")		
3110	Doors & Windows - Nailed to Concrete or Masonry	EA	76.75
3120	Bolted to Concrete or Masonry	"	84.75
3130	Doors & Windows - Nailed to Concrete or Masonry	BF	3.73
3140	Bolted to Concrete or Masonry	"	5.15
3150	Roof Edges - Nailed to Wood	"	2.98
3160	Bolted to Concrete (Incl. bolts)	"	4.46
3200	Grounds and Furring		
3210	1" x 6" Fastened to Wood	LF	2.37
3220	1" x 4"	"	1.77
3230	1" x 3"	"	1.60
3240	1" x 3" Fastened to Concrete/Masonry - Nailed	"	2.47
3250	Gun Driven	"	2.25
3260	PreClipped	"	2.80
3270	2" x 2" Suspended Framing	"	2.47
3280	Not Suspended Framing	"	2.14
3300	Grounds and Furring		
3310	1" x 6" Fastened to Wood	BF	4.76
3320	1" x 4"	"	5.61
3330	1" x 3"	"	6.29
3340	1" x 3" Fastened to Concrete/Masonry - Nailed	"	9.40
3400	Gun Driven	"	8.93
3410	PreClipped	"	10.96
3420	2" x 2" Suspended Framing	"	7.34
3430	Not Suspended Framing	"	6.34
3440	Cant Strips		
3450	4" x 4" Treated and Nailed	SF	2.81

		UNIT	COST
06999.10	**ROUGH CARPENTRY, Cont'd...**		
3460	Treated and Bolted (Including Bolts)	SF	4.07
3470	6" x 6" Treated and Nailed	"	5.64
3480	Treated and Bolted (Including Bolts)	"	7.16
3600	Building Papers and Sealers		
3610	15" Felt	SF	0.38
3620	Polyethylene - 4 mil	"	0.32
3630	6 mil	"	0.36
3640	Sill Sealer	"	1.40
06999.20	**FINISH CARPENTRY**		
1000	FINISH SIDINGS AND FACING MATERIALS (Exterior)		
1100	Boards and Beveled Sidings		
1200	Cedar Beveled - Clear Heart 1/2" x 4"	SF	8.65
1210	1/2" x 6"	"	7.62
1220	1/2" x 8"	"	6.52
1300	Rough Sawn 7/8" x 8"	"	6.64
1310	7/8" x 10"	"	6.86
1320	7/8" x 12"	"	6.86
1330	3/4" x 12"	"	5.97
1400	Redwood Beveled - Clear Heart		
1410	1/2" x 6"	SF	8.30
1430	5/8" x 10"	"	8.45
1440	3/4" x 6"	"	8.90
1450	3/4" x 8"	"	9.72
1500	3/4" x 6" Board - Clear	"	5.00
1510	3/4" Tongue & Groove	"	8.02
1520	1" Rustic Beveled - 6" and 8"	"	8.68
1530	5/4" Rustic Beveled - 6" and 8"	"	8.68
1540	Add for Metal Corners	EA	2.41
1550	Add for Mitering Corners	"	5.07
1600	Plywood Fir AC Smooth - One Side 1/4"	SF	3.03
1610	3/8"	"	3.23
1620	1/2"	"	3.28
1630	5/8"	"	3.47
1640	3/4"	"	3.89
1650	5/8" - Grooved and Rough Faced	"	4.00
1700	Cedar - Rough Sawn 3/8"	"	3.89
1710	5/8"	"	4.00
1720	3/4"	"	4.74
1730	5/8" - Grooved and Rough Faced	"	4.74
1740	Add for Wood Batten Strips, 1" x 2" - 4' O.C.	"	0.63
1750	Add for Splines	"	0.63
1800	Hardboard - Paneling and Lap Siding (Primed)		
1810	3/8" Rough Textured Paneling - 4' x 8'	SF	3.20
1820	7/16" Grooved - 4' x 8'	"	3.04
1830	7/16" Stucco Board Paneling	"	3.35
1840	7/16" x 8" Lap Siding	"	3.84
1860	Add for Pre-Finishing	"	0.31
1900	Shingles		
1910	Wood - 16" Red Cedar - 12" to Weather - #1	SF	6.31
1920	#2	"	4.98
1930	#3	"	4.67

		UNIT	COST
06999.20	**FINISH CARPENTRY, Cont'd...**		
1940	Red Cedar Hand Splits - #1, 24" - 1/2" to 3/4"	SF	5.30
1950	24" - 3/4" to 1-1/4"	"	5.90
1960	#2	"	5.30
1970	#3	"	4.98
2000	Add for Fire Retardant	"	0.76
2010	Add for 3/8" Backer Board	"	1.26
2020	Add for Metal Corners	EA	2.44
2030	Add for Ridges, Hips and Corners	LF	6.51
2040	Add for 8" to Weather	PCT	29.50
2100	Facia - Pine #2 1" x 8"	LF	3.08
2110	Cedar #3 1" x 8"	"	4.88
2120	Redwood - Clear 1" x 8"	"	6.49
2130	Plywood 5/8" - ACX 1" x 8"	"	3.18
2140	Facia - Pine #2 1" x 8"	SF	4.61
2200	Cedar #3 1" x 8"	"	6.83
2210	Redwood - Clear 1" x 8"	"	9.93
2220	Plywood 5/8" - ACX 1" x 8"	"	4.69
2300	FINISH WALLS (Interior) (15% Added for Waste)	"	5.13
2400	Boards, Cedar - #3 1" x 6" and 1" x 8"	"	5.13
2410	Knotty 1" x 6" and 1" x 8"	"	5.45
2420	D Grade 1" x 6" and 1" x 8"	"	6.60
2430	Aromatic 1" x 6" and 1" x 8"	"	6.81
2440	Redwood - Construction 1" x 6" and 1" x 8"	"	8.16
2450	Clear 1" x 6" and 1" x 8"	"	8.45
2460	Fir - Beaded 5/8" x 4"	"	7.25
2470	Pine - #2 1" x 6" and 1" x 8"	"	9.15
2500	Hardboard (Paneling) Tempered - 1/8"	"	6.69
2510	1/4"	"	5.10
2600	Plywood (Prefinished Paneling) 1/4" Birch - Natural	"	2.13
2610	3/4" Birch - Natural	"	2.33
2620	1/4" Birch - White	"	4.45
2630	1/4" Oak - Rotary Cut	"	5.86
2640	3/4"	"	7.72
2700	1/4" Oak - White	"	5.66
2710	1/4" Mahogany (Lauan)	"	7.04
2720	3/4"	"	9.49
2730	3/4" Mahogany (African)	"	4.46
2740	1/4" Walnut	"	6.38
2800	Gypsum Board (Paneling)	"	7.71
2810	Prefinished	"	8.47
2820	Red Oak Plastic		
06999.30	**MILLWORK & CUSTOM WOODWORK**		
1000	CUSTOM CABINET WORK (Red Oak or Birch)		
1010	Base Cabinets - Avg. 35" H x 24" D with Drawer	LF	460
1020	Sink Fronts	"	178
1030	Corner Cabinets	"	370
1040	Add per Drawer	"	76.75
1050	Upper Cabinets - Avg. 30" High x 12" Deep	"	230
2000	Utility Cabinets - Avg. 84" High x 24" Deep	"	550
2010	China or Corner Cabinets - 84" High	EA	1,050
2020	Oven Cabinets - 84" High x 24" Deep	"	620

		UNIT	COST
06999.30	**MILLWORK & CUSTOM WOODWORK, Cont'd...**		
2030	Vanity Cabinets - Avg. 30" High x 21" Deep	EA	370
2040	Deduct for Prefinishing wood	PCT	10.81
2050	Base Cabinets - Avg. 35" H x 24" D with Drawer	EA	280
2060	Sink Fronts	LF	178
2070	Corner Cabinets	"	400
2080	Add per Drawer	"	78.25
2090	Upper Cabinets - Avg. 30" High x 12" Deep	"	220
3000	Utility Cabinets - Avg. 84" High x 24" Deep	"	550
3010	China or Corner Cabinets - 84" High	"	1,060
3020	Oven Cabinets - 84" High x 24" Deep	"	620
3030	Vanity Cabinets - Avg. 30" High x 21" Deep	"	360
4000	COUNTERTOPS - 25"		
4010	Plastic Laminated with 4" Back Splash	LF	106
4020	Deduct for No Back Splash	"	17.66
4030	Granite - 1-1/4" - Artificial	"	188
4040	3/4" - Artificial	"	148
4050	Marble	"	179
4070	Wood Cutting Block	"	179
5000	Stainless Steel	"	260
5010	Polyester-Acrylic Solid Surface	"	290
5020	Quartz	"	162
5030	Plastic (Polymer)	"	111
6000	CUSTOM DOOR FRAMES - Including 2 Sides Trim		
6010	Birch	EA	470
6020	Fir	"	280
6030	Poplar	"	280
6040	Oak	"	520
6050	Pine	"	390
6060	Walnut	"	620
8000	MOULDINGS AND TRIM		
8010	Apron 7/16" x 2"	LF	3.60
8020	Astragal 1-3/4" x 2-1/4"	"	8.55
8030	Base 7/16" x 2-3/4"	"	5.17
8050	Base Shoe 7/16" x 2-3/4"	"	4.14
8060	Batten Strip 5/8" x 1-5/8"	"	3.56
8070	Brick Mould 1-1/14" x 2"	"	4.15
9080	Casing 1-1/16" x 2-1/4"	"	4.50
9100	Chair Rail 5/8" x 1-3/4"	"	4.71
9110	Closet Rod 1-5/16"	"	4.92
9120	Corner Bead 1-1/8" x 1-1/8"	"	6.22
9130	Cove Moulding 3/4" x 3/4"	"	3.87
9140	Crown Moulding 9/16" x 3-5/8"	"	5.77
9160	Drip Cap	"	4.65
9170	Half Round 1/2" x 1"	"	4.98
9180	Hand Rail 1 5/8" x 1-3/4"	"	6.94
9190	Hook Strip 5/8" x 2-1/2"	"	4.00
9200	Picture Mould 3/4" x 1-1/2"	"	4.44
9210	Quarter Round 3/4" x 3/4"	"	3.16
9220	1/2" x 1/2"	"	2.43
9230	Sill 3/4" x 2-1/2"	"	5.30
9240	Stool 11/16 x 2-1/2"	"	6.37
9250	BIRCH		

		UNIT	COST
06999.30	**MILLWORK & CUSTOM WOODWORK, Cont'd...**		
9260	STAIRS (Treads, Risers, Skirt Boards)	LF	60.50
9270	SHELVING (12" Deep)	SF	64.25
9280	CUSTOM PANELING	"	43.75
9290	OAK		
9300	STAIRS (Treads, Risers, Skirt Boards)	LF	47.25
9310	SHELVING (12" Deep)	SF	57.25
9320	CUSTOM PANELING	"	32.75
9330	THRESHOLDS - 3/4" x 3-1/2"	LF	31.00
9340	PINE		
9350	STAIRS (Treads, Risers, Skirt Boards)	LF	35.50
9360	SHELVING (12" Deep)	"	39.50
9370	CUSTOM PANELING	SF	27.00
06999.40	**GLUE LAMINATE**		
0010	Arches 80' Span	SF	15.65
0020	Beams & Purlins 40' Span	"	10.84
0030	Deck Fir - 3" x 6"	"	12.92
0040	4" x 6"	"	13.90
0050	Deck Cedar - 3" x 6"	"	16.71
0060	4" x 6"	"	18.82
0070	Deck Pine - 3" x 6"	"	11.34
06999.50	**PREFABRICATED WOOD COMPONENTS**		
1000	WOOD TRUSSED RAFTERS		
1010	24" O.C. - 4-12 Pitch		
2000	Span To 16' w/ Supports	EA	210
2010	20'	"	220
2030	24'	"	260
2050	28'	"	270
2070	32'	"	330
3000	Add for 5-12 Pitch	PCT	5.21
3010	Add for 6-12 Pitch	"	10.42
3020	Add for Scissor Truss	EA	46.75
3030	Add for Gable End 24'	"	101
3040	Span To 16' w/ Supports	SF	6.62
3050	20'	"	5.36
3070	24'	"	5.07
3100	28'	"	5.11
3120	32'	"	5.72
5000	Add for 5-12 Pitch	PCT	15.63
5010	Add for 6-12 Pitch	"	26.05
5020	Add for Scissor Truss	SF	36.75
5030	Add for Gable End 24'	"	0.67
6000	WOOD FLOOR TRUSS JOISTS (TJI)		
6010	Open Web - Wood or Metal - 24" O.C.		
6020	Up To 23' Span x 12" Single	LF	13.24
6030	To 24' Span x 15" Cord	"	13.58
6050	To 30' Span x 21"	"	14.79
6060	To 25' Span x 15" Double	"	12.89
6070	To 30' Span x 18" Cord	"	13.65
6080	To 36' Span x 21"	"	15.24
7000	Plywood Web - 24" O.C.		
7010	Up To 15' x 9-1/2"	LF	8.57

		UNIT	COST
06999.50	**PREFABRICATED WOOD COMPONENTS, Cont'd...**		
7030	To 21' x 14"	LF	9.51
7040	To 22' x 16"	"	10.74
8000	Open Web - Wood or Metal - 24" O.C.		
8010	Up To 23' Span x 12" Single	SF	6.60
8020	To 24' Span x 15" Cord	"	5.39
8030	To 27' Span x 18"	"	7.01
8040	To 30' Span x 21"	"	7.44
8050	To 25' Span x 15" Double	"	6.51
8060	To 30' Span x 18" Cord	"	6.65
8070	To 36' Span x 21"	"	7.52
8500	Plywood Web - 24" O.C.		
8510	Up To 15' x 9-1/2"	SF	4.36
8520	To 19' x 11-7/8"	"	4.36
8530	To 21' x 14"	"	4.87
8540	To 22' x 16"	"	5.36
06999.51	**LAMINATED VENEER STRUCTURAL BEAMS**		
0005	Micro Lam - Plywood - 24" O.C.		
0010	9-1/2" x 1-3/4"	LF	12.94
0020	11-7/8" x 1-3/4"	"	14.02
0030	14" x 1-3/4"	"	16.64
0040	16" x 1-3/4"	"	18.39
1000	Glue Lam- Dimension Lumber- 24" O.C.		
1010	9" x 3-1/2"	LF	24.50
1020	12" x 3-1/2"	"	30.00
1040	9" x 5-1/2"	"	35.75
1050	12" x 5-1/2"	"	43.50
1070	18" x 5-1/2"	"	67.00
2000	Add for Architectural Grade	PCT	26.05
2010	Add for 6-3/4"	"	41.68
3000	Micro Lam - Plywood - 24" O.C.		
3010	9-1/2" x 1-3/4"	SF	6.53
3020	11-7/8" x 1-3/4"	"	7.01
3040	16" x 1-3/4"	"	9.12
4000	Glue Lam - Dimension Lumber - 24" O.C.		
4010	9" x 3-1/2"	SF	12.08
4020	12" x 3-1/2"	"	14.94
4040	9" x 5-1/2"	"	17.91
4050	12" x 5-1/2"	"	21.91
4070	18" x 5-1/2"	"	33.25
06999.60	**WOOD TREATMENTS**		
0010	PRESERVATIVES – PRESSURE TREATED		
0020	Dimensions	BF	0.34
0030	Timbers	"	0.50
0040	FIRE RETARDENTS		
0050	Dimensions & Timbers	BF	0.65
06999.70	**ROUGH HARDWARE**		
1000	NAILS	BF	0.03
2000	JOIST HANGERS	EA	4.25
3000	BOLTS - 5/8" x 12"	"	5.53

DCD

Design Cost Data

TABLE OF CONTENTS   PAGE

		UNIT	LABOR	MAT.	TOTAL
07100.10	**WATERPROOFING**				
0100	Membrane waterproofing, elastomeric				
1020	Butyl				
1040	1/32" thick	SF	2.04	1.47	3.51
1060	1/16" thick	"	2.12	1.91	4.03
1140	Neoprene				
1160	1/32" thick	SF	2.04	2.51	4.55
1180	1/16" thick	"	2.12	3.60	5.72
1260	Plastic vapor barrier (polyethylene)				
1280	4 mil	SF	0.20	0.05	0.25
1300	6 mil	"	0.20	0.09	0.29
1320	10 mil	"	0.25	0.12	0.37
1400	Bituminous membrane, asphalt felt, 15 lb.				
1440	One ply	SF	1.27	0.87	2.14
1460	Two ply	"	1.54	1.03	2.57
1480	Three ply	"	1.82	1.27	3.09
07160.10	**BITUMINOUS DAMPPROOFING**				
0100	Building paper, asphalt felt				
0120	15 lb	SF	2.04	0.19	2.23
0140	30 lb	"	2.12	0.37	2.49
1000	Asphalt, troweled, cold, primer plus				
1020	1 coat	SF	1.70	0.68	2.38
1040	2 coats	"	2.55	1.43	3.98
1060	3 coats	"	3.19	2.04	5.23
1200	Fibrous asphalt, hot troweled, primer plus				
1220	1 coat	SF	2.04	0.68	2.72
1240	2 coats	"	2.83	1.43	4.26
1260	3 coats	"	3.64	2.04	5.68
07190.10	**VAPOR BARRIERS**				
0980	Vapor barrier, polyethylene				
1000	2 mil	SF	0.25	0.02	0.27
1010	6 mil	"	0.25	0.07	0.32
1020	8 mil	"	0.28	0.08	0.36
1040	10 mil	"	0.28	0.09	0.37
07210.10	**BATT INSULATION**				
0980	Ceiling, fiberglass, unfaced				
1000	3-1/2" thick, R11	SF	0.60	0.42	1.02
1020	6" thick, R19	"	0.68	0.56	1.24
1030	9" thick, R30	"	0.78	1.10	1.88
1035	Suspended ceiling, unfaced				
1040	3-1/2" thick, R11	SF	0.56	0.42	0.98
1060	6" thick, R19	"	0.63	0.56	1.19
1070	9" thick, R30	"	0.72	1.10	1.82
1075	Crawl space, unfaced				
1080	3-1/2" thick, R11	SF	0.78	0.42	1.20
1100	6" thick, R19	"	0.85	0.56	1.41
1120	9" thick, R30	"	0.92	1.10	2.02
2000	Wall, fiberglass				
2010	Paper backed				
2020	2" thick, R7	SF	0.53	0.33	0.86
2040	3" thick, R8	"	0.56	0.36	0.92
2060	4" thick, R11	"	0.60	0.59	1.19

		UNIT	LABOR	MAT.	TOTAL
07210.10	**BATT INSULATION, Cont'd...**				
2080	6" thick, R19	SF	0.63	0.88	1.51
2090	Foil backed, 1 side				
2100	2" thick, R7	SF	0.53	0.63	1.16
2120	3" thick, R11	"	0.56	0.68	1.24
2140	4" thick, R14	"	0.60	0.71	1.31
2160	6" thick, R21	"	0.63	0.93	1.56
07210.20	**BOARD INSULATION**				
2200	Perlite board, roof				
2220	1.00" thick, R2.78	SF	0.42	0.57	0.99
2240	1.50" thick, R4.17	"	0.44	0.90	1.34
2580	Rigid urethane				
2600	1" thick, R6.67	SF	0.42	1.09	1.51
2640	1.50" thick, R11.11	"	0.44	1.49	1.93
2780	Polystyrene				
2800	1.0" thick, R4.17	SF	0.42	0.41	0.83
2820	1.5" thick, R6.26	"	0.44	0.63	1.07
07210.60	**LOOSE FILL INSULATION**				
1000	Blown-in type				
1010	Fiberglass				
1020	5" thick, R11	SF	0.42	0.41	0.83
1040	6" thick, R13	"	0.51	0.48	0.99
1060	9" thick, R19	"	0.72	0.58	1.30
07210.70	**SPRAYED INSULATION**				
1000	Foam, sprayed on				
1010	Polystyrene				
1020	1" thick, R4	SF	0.51	0.69	1.20
1040	2" thick, R8	"	0.68	1.34	2.02
1050	Urethane				
1060	1" thick, R4	SF	0.51	0.65	1.16
1080	2" thick, R8	"	0.68	1.24	1.92
07310.10	**ASPHALT SHINGLES**				
1000	Standard asphalt shingles, strip shingles				
1020	210 lb/square	SQ	62.00	90.00	152
1040	235 lb/square	"	69.00	95.00	164
1060	240 lb/square	"	78.00	99.00	177
1080	260 lb/square	"	89.00	140	229
1100	300 lb/square	"	100	150	250
1120	385 lb/square	"	120	210	330
5980	Roll roofing, mineral surface				
6000	90 lb	SQ	44.50	55.00	99.50
6020	110 lb	"	52.00	92.00	144
6040	140 lb	"	62.00	95.00	157
07310.50	**METAL SHINGLES**				
0980	Aluminum, .020" thick				
1000	Plain	SQ	120	280	400
1020	Colors	"	120	310	430
1960	Steel, galvanized				
1980	26 ga.				
2000	Plain	SQ	120	350	470
2020	Colors	"	120	440	560
2030	24 ga.				

DIVISION # 07 THERMAL AND MOISTURE

		UNIT	LABOR	MAT.	TOTAL
07310.50	**METAL SHINGLES, Cont'd...**				
2040	Plain	SQ	120	400	520
2060	Colors	"	120	510	630
07310.60	**SLATE SHINGLES**				
0960	Slate shingles				
0980	Pennsylvania				
1000	Ribbon	SQ	310	600	910
1020	Clear	"	310	770	1,080
1030	Vermont				
1040	Black	SQ	310	710	1,020
1060	Gray	"	310	780	1,090
1070	Green	"	310	800	1,110
1080	Red	"	310	1,440	1,750
07310.70	**WOOD SHINGLES**				
1000	Wood shingles, on roofs				
1010	White cedar, #1 shingles				
1020	4" exposure	SQ	210	270	480
1040	5" exposure	"	160	240	400
1050	#2 shingles				
1060	4" exposure	SQ	210	190	400
1080	5" exposure	"	160	160	320
1090	Resquared and rebutted				
1100	4" exposure	SQ	210	240	450
1120	5" exposure	"	160	200	360
1140	On walls				
1150	White cedar, #1 shingles				
1160	4" exposure	SQ	310	270	580
1180	5" exposure	"	250	240	490
1200	6" exposure	"	210	200	410
1210	#2 shingles				
1220	4" exposure	SQ	310	190	500
1240	5" exposure	"	250	160	410
1260	6" exposure	"	210	130	340
1300	Add for fire retarding	"			120
07310.80	**WOOD SHAKES**				
2010	Shakes, hand split, 24" red cedar, on roofs				
2020	5" exposure	SQ	310	300	610
2040	7" exposure	"	250	280	530
2060	9" exposure	"	210	260	470
2080	On walls				
2100	6" exposure	SQ	310	280	590
2120	8" exposure	"	250	270	520
2140	10" exposure	"	210	260	470
3000	Add for fire retarding	"			78.00
07310.90	**CLAY ROOF TILE**				
1010	Spanish tile, terracota	SQ	160	420	580
1020	Mission tile, terracota	"	210	820	1,030
1030	French tile, rustic blue, green patina	"	180	950	1,130
07460.10	**METAL SIDING PANELS**				
1000	Aluminum siding panels				
1020	Corrugated				
1030	Plain finish				

		UNIT	LABOR	MAT.	TOTAL
07460.10	**METAL SIDING PANELS, Cont'd...**				
1040	.024"	SF	2.88	2.23	5.11
1060	.032"	"	2.88	2.62	5.50
1070	Painted finish				
1080	.024"	SF	2.88	2.78	5.66
1100	.032"	"	2.88	3.19	6.07
2000	Steel siding panels				
2040	Corrugated				
2080	22 ga.	SF	4.80	2.73	7.53
2100	24 ga.	"	4.80	2.49	7.29
07460.50	**PLASTIC SIDING**				
1000	Horizontal vinyl siding, solid				
1010	8" wide				
1020	Standard	SF	2.50	1.35	3.85
1040	Insulated	"	2.50	1.64	4.14
1050	10" wide				
1060	Standard	SF	2.33	1.40	3.73
1080	Insulated	"	2.33	1.68	4.01
8500	Vinyl moldings for doors and windows	LF	2.60	0.87	3.47
07460.60	**PLYWOOD SIDING**				
1000	Rough sawn cedar, 3/8" thick	SF	2.17	2.16	4.33
1020	Fir, 3/8" thick	"	2.17	1.19	3.36
1980	Texture 1-11, 5/8" thick				
2000	Cedar	SF	2.33	2.92	5.25
2020	Fir	"	2.33	2.04	4.37
2040	Redwood	"	2.22	3.14	5.36
2060	Southern Yellow Pine	"	2.33	1.66	3.99
07460.80	**WOOD SIDING**				
1000	Beveled siding, cedar				
1010	A grade				
1040	1/2 x 8	SF	2.60	5.00	7.60
1060	3/4 x 10	"	2.17	6.43	8.60
1070	Clear				
1100	1/2 x 8	SF	2.60	5.56	8.16
1120	3/4 x 10	"	2.17	7.45	9.62
1130	B grade				
1160	1/2 x 8	SF	2.60	5.94	8.54
1180	3/4 x 10	"	2.17	5.60	7.77
2000	Board and batten				
2010	Cedar				
2020	1x6	SF	3.26	6.84	10.10
2040	1x8	"	2.60	6.22	8.82
2060	1x10	"	2.33	5.62	7.95
2080	1x12	"	2.10	5.04	7.14
2090	Pine				
2100	1x6	SF	3.26	1.73	4.99
2120	1x8	"	2.60	1.69	4.29
2140	1x10	"	2.33	1.62	3.95
2160	1x12	"	2.10	1.49	3.59
3000	Tongue and groove				
3010	Cedar				
3020	1x4	SF	3.62	6.43	10.05

		UNIT	LABOR	MAT.	TOTAL
07460.80	**WOOD SIDING, Cont'd...**				
3040	1x6	SF	3.43	6.18	9.61
3060	1x8	"	3.26	5.80	9.06
3080	1x10	"	3.10	5.69	8.79
3090	Pine				
3100	1x4	SF	3.62	1.93	5.55
3120	1x6	"	3.43	1.82	5.25
3140	1x8	"	3.26	1.70	4.96
3160	1x10	"	3.10	1.62	4.72
07510.10	**BUILT-UP ASPHALT ROOFING**				
0980	Built-up roofing, asphalt felt, including gravel				
1000	2 ply	SQ	160	88.00	248
1500	3 ply	"	210	120	330
2000	4 ply	"	250	170	420
2195	Cant strip, 4" x 4"				
2200	Treated wood	LF	1.77	2.56	4.33
2260	Foamglass	"	1.55	2.20	3.75
8000	New gravel for built-up roofing, 400 lb/sq	SQ	120	44.50	165
07530.10	**SINGLE-PLY ROOFING**				
2000	Elastic sheet roofing				
2060	Neoprene, 1/16" thick	SF	0.77	2.83	3.60
2115	PVC				
2120	45 mil	SF	0.77	2.34	3.11
2200	Flashing				
2220	Pipe flashing, 90 mil thick				
2260	1" pipe	EA	15.50	34.00	49.50
2360	Neoprene flashing, 60 mil thick strip				
2380	6" wide	LF	5.19	1.72	6.91
2390	12" wide	"	7.78	3.38	11.16
07610.10	**METAL ROOFING**				
1000	Sheet metal roofing, copper, 16 oz, batten seam	SQ	420	1,800	2,220
1020	Standing seam	"	390	1,760	2,150
2000	Aluminum roofing, natural finish				
2005	Corrugated, on steel frame				
2010	.0175" thick	SQ	180	140	320
2040	.0215" thick	"	180	180	360
2060	.024" thick	"	180	210	390
2080	.032" thick	"	180	260	440
2100	V-beam, on steel frame				
2120	.032" thick	SQ	180	270	450
2130	.040" thick	"	180	290	470
2140	.050" thick	"	180	370	550
2200	Ridge cap				
2220	.019" thick	LF	2.07	4.25	6.32
2500	Corrugated galvanized steel roofing, on steel frame				
2520	28 ga.	SQ	180	230	410
2540	26 ga.	"	180	270	450
2550	24 ga.	"	180	300	480
2560	22 ga.	"	180	330	510

		UNIT	LABOR	MAT.	TOTAL
07620.10	**FLASHING AND TRIM**				
0050	Counter flashing				
0060	Aluminum, .032"	SF	6.22	2.09	8.31
0100	Stainless steel, .015"	"	6.22	6.69	12.91
0105	Copper				
0110	16 oz.	SF	6.22	9.36	15.58
0112	20 oz.	"	6.22	11.00	17.22
0114	24 oz.	"	6.22	13.50	19.72
0116	32 oz.	"	6.22	16.50	22.72
0118	Valley flashing				
0120	Aluminum, .032"	SF	3.89	1.74	5.63
0130	Stainless steel, .015	"	3.89	5.56	9.45
0135	Copper				
0140	16 oz.	SF	3.89	9.36	13.25
0160	20 oz.	"	5.19	11.00	16.19
0180	24 oz.	"	3.89	13.50	17.39
0200	32 oz.	"	3.89	16.50	20.39
0380	Base flashing				
0400	Aluminum, .040"	SF	5.19	2.60	7.79
0410	Stainless steel, .018"	"	5.19	6.65	11.84
0415	Copper				
0420	16 oz.	SF	5.19	9.36	14.55
0422	20 oz.	"	3.89	11.00	14.89
0424	24 oz.	"	5.19	13.50	18.69
0426	32 oz.	"	5.19	16.50	21.69
07620.20	**GUTTERS AND DOWNSPOUTS**				
1500	Copper gutter and downspout				
1520	Downspouts, 16 oz. copper				
1530	Round				
1540	3" dia.	LF	4.15	12.50	16.65
1550	4" dia.	"	4.15	15.50	19.65
1800	Gutters, 16 oz. copper				
1810	Half round				
1820	4" wide	LF	6.22	11.25	17.47
1840	5" wide	"	6.92	13.75	20.67
1860	Type K				
1880	4" wide	LF	6.22	12.50	18.72
1890	5" wide	"	6.92	13.00	19.92
3000	Aluminum gutter and downspout				
3005	Downspouts				
3010	2" x 3"	LF	4.15	1.45	5.60
3030	3" x 4"	"	4.44	2.00	6.44
3035	4" x 5"	"	4.79	2.31	7.10
3038	Round				
3040	3" dia.	LF	4.15	2.44	6.59
3050	4" dia.	"	4.44	3.13	7.57
3240	Gutters, stock units				
3260	4" wide	LF	6.55	2.26	8.81
3270	5" wide	"	6.92	2.69	9.61
4101	Galvanized steel gutter and downspout				
4111	Downspouts, round corrugated				
4121	3" dia.	LF	4.15	2.09	6.24

		UNIT	LABOR	MAT.	TOTAL
07620.20	**GUTTERS AND DOWNSPOUTS, Cont'd...**				
4131	4" dia.	LF	4.15	2.80	6.95
4141	5" dia.	"	4.44	4.18	8.62
4151	6" dia.	"	4.44	5.54	9.98
4161	Rectangular				
4171	2" x 3"	LF	4.15	1.89	6.04
4191	3" x 4"	"	3.89	2.70	6.59
4201	4" x 4"	"	3.89	3.38	7.27
4300	Gutters, stock units				
4310	5" wide				
4320	Plain	LF	6.92	1.82	8.74
4330	Painted	"	6.92	1.98	8.90
4335	6" wide				
4340	Plain	LF	7.32	2.55	9.87
4360	Painted	"	7.32	2.86	10.18
07810.10	**PLASTIC SKYLIGHTS**				
1030	Single thickness, not including mounting curb				
1040	2' x 4'	EA	78.00	410	488
1050	4' x 4'	"	100	550	650
1060	5' x 5'	"	160	730	890
1070	6' x 8'	"	210	1,560	1,770
07920.10	**CAULKING**				
0100	Caulk exterior, two component				
0120	1/4 x 1/2	LF	3.26	0.43	3.69
0140	3/8 x 1/2	"	3.62	0.66	4.28
0160	1/2 x 1/2	"	4.07	0.90	4.97
0220	Caulk interior, single component				
0240	1/4 x 1/2	LF	3.10	0.29	3.39
0260	3/8 x 1/2	"	3.43	0.41	3.84
0280	1/2 x 1/2	"	3.83	0.54	4.37

		UNIT	COST
07999.10	**WATERPROOFING**		
0010	1-Ply Membrane Felt 15#	SF	2.06
0020	2-Ply Membrane Felt 15#	"	3.13
0030	Hydrolithic	"	3.38
0040	Elastomeric Rubberized Asphalt with Poly Sheet	"	3.13
0060	Metallic Oxide 3-Coat	"	5.35
0070	Vinyl Plastic	"	3.75
0080	Bentonite 3/8" - Trowel	"	3.66
0090	5/8" - Panels	"	3.76
07999.20	**DAMPPROOFING**		
0010	Asphalt Trowel Mastic 1/16"	SF	1.26
0020	1/8"	"	1.69
0030	Spray Liquid 1-Coat	"	0.78
0040	2-Coat	"	1.10
0050	Brush Liquid 2-Coat	"	1.24
1000	Hot Mop 1-Coat and Primer	"	2.06
1010	1 Fibrous Asphalt	"	2.25
1020	Cementitious Per Coat 1/2" Coat	"	1.32
1030	Silicone 1-Coat	"	0.86
1040	2-Coat	"	1.51
1050	Add for Scaffold and Lift Operations	"	0.77
07999.30	**BUILDING INSULATION**		
1000	FLEXIBLE		
1010	Fiberglass 2-1/4" R 7.40	SF	0.97
1020	3-1/2" R 11.00	"	1.14
1040	6" R 19.00	"	1.44
1060	12" R 38.00	"	1.99
1100	Add for Ceiling Work	"	0.12
1110	Add for Paper Faced	"	0.15
1120	Add for Polystyrene Barrier (2m)	"	0.15
1130	Add for Scaffold Work	"	0.98
1200	RIGID		
1300	Fiberglass 1" R 4.35 3# Density	SF	1.51
1310	1-1/2" R 6.52 3# Density	"	1.94
1320	2" R 8.70 3# Density	"	2.29
1400	Styrene, Molded 1" R 4.3.5	"	7.03
1410	1-1/2" R 6.52	"	1.29
1420	2" R 7.69	"	1.72
1430	2" T&G R 7.69	"	2.21
1500	Styrene, Extruded 1" R 5.40	"	1.57
1510	1-1/2" R 6.52	"	1.86
1520	2" R 7.69	"	1.99
1530	2" T&G R 7.69	"	2.21
1600	Perlite 1" R 2.78	"	1.57
1610	2" R 5.56	"	2.51
1700	Urethane 1" R 6.67	"	1.86
1710	2" R 13.34	"	2.36
1720	Add for Glued Applications	"	0.12
1800	LOOSE 1" Styrene R 3.8	"	1.10
1810	1" Fiberglass R 2.2	"	1.06
1820	1" Rock Wool R 2.9	"	1.09
1830	1" Cellulose R 3.7	"	1.03

		UNIT	COST
07999.30	**BUILDING INSULATION, Cont'd...**		
1900	FOAMED Urethane Per Inch	SF	2.68
2000	SPRAYED Cellulose Per Inch	"	1.81
2100	Polystyrene	"	2.17
2200	Urethane	"	3.07
2300	ALUMINUM PAPER	"	0.82
2400	VINYL FACED FIBERGLASS	"	2.46
07999.80	**SHINGLE ROOFING**		
1000	ASPHALT SHINGLES 235# Seal Down	SF	1.73
1010	300# Laminated	"	1.96
1020	325# Fire Resistant	"	2.83
1030	Timberline	"	2.76
1040	340# Tab Lock	"	3.26
2000	ASPHALT ROLL 90#	"	0.95
3000	FIBERGLASS SHINGLES 215#	"	1.67
3010	250#	"	1.89
4000	RED CEDAR 16" x 5" to Weather #1 Grade	"	4.89
4010	#2 Grade	"	4.36
4020	#3 Grade	"	3.11
4030	24" x 10" to Weather - Hand Splits - 1/2" x 3/4"	"	15.83
4040	3/4" x 1-1/4"	"	15.96
4050	Add for Fire Retardant	"	1.24
5000	METAL Aluminum 020 mil	"	4.90
6000	Anodized 020 Mil	"	5.66
7000	Steel, Enameled Colored	"	7.28
8000	Galvanized Colored	"	6.29
9000	Add to Above for 15# Felt Underlayment		
9010	Add for Base Starter	SF	0.90
9020	Add for Ice & Water Starter	"	1.50
9030	Add for Boston Ridge	"	3.40
9040	Add for Removal & Haul Away	"	1.30
9050	Add for Pitches Over 5-12 each Pitch Increase	"	0.07
9060	Add for Chimneys, Skylights and Bay Windows	EA	98.50
07999.90	**BUILT-UP ROOFING**		
1000	MEMBRANE		
1010	3-Ply Asphalt & Gravel - R 10.0	SQ	850
1020	R 16.6	"	870
1030	4-Ply - R 10.0	"	880
1040	R 16.6	"	890
1050	5-Ply - R 10.0	"	910
1060	R 16.6	"	950
2000	Add for Sheet Rock over Steel Deck - 5/8"	"	177
2010	Add for Fiberglass Insulation	"	61.00
2020	Add for Pitch and Gravel	"	93.75
2030	Add for Sloped Roofs	"	65.75
2040	Add for Thermal Barrier	"	50.75
2050	Add for Upside Down Roofing System	"	177
3000	SINGLE-PLY - 60M Butylene Roofing & Gravel Ballast - R 10.0	"	610
3010	Mech. Fastened - R 10.0	"	590
3020	PVC & EPDM Roofing & Gravel Ballast - R 10.0	"	560
3030	Mech. Fastened - R 10.0	"	590
3040	Blocking and Cants Not Included		

TABLE OF CONTENTS   PAGE

DIVISION # 08 DOORS AND WINDOWS

		UNIT	LABOR	MAT.	TOTAL
08110.10	**METAL DOORS**				
1000	Flush hollow metal, std. duty, 20 ga., 1-3/8" thick				
1020	2-6 x 6-8	EA	72.00	350	422
1080	3-0 x 6-8	"	72.00	420	492
1090	1-3/4" thick				
1100	2-6 x 6-8	EA	72.00	410	482
1150	3-0 x 6-8	"	72.00	470	542
1200	2-6 x 7-0	"	72.00	450	522
1240	3-0 x 7-0	"	72.00	500	572
2110	Heavy duty, 20 ga., unrated, 1-3/4"				
2130	2-8 x 6-8	EA	72.00	450	522
2135	3-0 x 6-8	"	72.00	490	562
2140	2-8 x 7-0	"	72.00	520	592
2150	3-0 x 7-0	"	72.00	500	572
2200	18 ga., 1-3/4", unrated door				
2210	2-0 x 7-0	EA	72.00	480	552
2235	2-6 x 7-0	"	72.00	480	552
2260	3-0 x 7-0	"	72.00	540	612
2270	3-4 x 7-0	"	72.00	560	632
2310	2", unrated door				
2320	2-0 x 7-0	EA	82.00	530	612
2340	2-6 x 7-0	"	82.00	530	612
2360	3-0 x 7-0	"	82.00	600	682
2370	3-4 x 7-0	"	82.00	610	692
2400	Galvanized metal door				
2410	3-0 x 7-0	EA	82.00	620	702
2450	For lead lining in doors	"			1,120
2460	For sound attenuation	"			100
4280	Vision glass				
4300	8" x 8"	EA	82.00	130	212
4320	8" x 48"	"	82.00	200	282
4340	Fixed metal louver	"	65.00	290	355
4350	For fire rating, add				
4370	3 hr door	EA			490
4380	1-1/2 hr door	"			220
4400	3/4 hr door	"			110
4430	1' extra height, add to material, 20%				
4440	1'6" extra height, add to material, 60%				
4470	For dutch doors with shelf, add to material, 100%				
5000	Stainless steel, general application				
5010	3'x7'	EA	720	2,150	2,870
5020	5'x7'	"	960	3,110	4,070
6000	Heavy impact, s-core, 18 ga., stainless				
6010	3'x7'	EA	720	2,270	2,990
6020	5'x7'	"	960	3,780	4,740
08110.40	**METAL DOOR FRAMES**				
1000	Hollow metal, stock, 18 ga., 4-3/4" x 1-3/4"				
1020	2-0 x 7-0	EA	82.00	160	242
1060	2-6 x 7-0	"	82.00	190	272
1100	3-0 x 7-0	"	82.00	190	272
1120	4-0 x 7-0	"	110	210	320
1140	5-0 x 7-0	"	110	220	330

© 2020 By Design & Construction Resources

		UNIT	LABOR	MAT.	TOTAL
08110.40	**METAL DOOR FRAMES, Cont'd...**				
1160	6-0 x 7-0	EA	110	260	370
1500	16 ga., 6-3/4" x 1-3/4"				
1520	2-0 x 7-0	EA	90.00	190	280
1535	2-6 x 7-0	"	90.00	180	270
1550	3-0 x 7-0	"	90.00	200	290
1560	4-0 x 7-0	"	120	230	350
1580	6-0 x 7-0	"	120	260	380
08210.10	**WOOD DOORS**				
0980	Solid core, 1-3/8" thick				
1000	Birch faced				
1020	2-4 x 7-0	EA	82.00	180	262
1060	3-0 x 7-0	"	82.00	180	262
1070	3-4 x 7-0	"	82.00	370	452
1080	2-4 x 6-8	"	82.00	180	262
1090	2-6 x 6-8	"	82.00	180	262
1100	3-0 x 6-8	"	82.00	180	262
1120	Lauan faced				
1140	2-4 x 6-8	EA	82.00	160	242
1180	3-0 x 6-8	"	82.00	180	262
1200	3-4 x 6-8	"	82.00	190	272
1300	Tempered hardboard faced				
1320	2-4 x 7-0	EA	82.00	200	282
1360	3-0 x 7-0	"	82.00	240	322
1380	3-4 x 7-0	"	82.00	250	332
1420	Hollow core, 1-3/8" thick				
1440	Birch faced				
1460	2-4 x 7-0	EA	82.00	160	242
1500	3-0 x 7-0	"	82.00	170	252
1520	3-4 x 7-0	"	82.00	180	262
1600	Lauan faced				
1620	2-4 x 6-8	EA	82.00	69.00	151
1630	2-6 x 6-8	"	82.00	74.00	156
1660	3-0 x 6-8	"	82.00	97.00	179
1680	3-4 x 6-8	"	82.00	110	192
1740	Tempered hardboard faced				
1770	2-6 x 7-0	EA	82.00	90.00	172
1800	3-0 x 7-0	"	82.00	110	192
1820	3-4 x 7-0	"	82.00	120	202
1900	Solid core, 1-3/4" thick				
1920	Birch faced				
1940	2-4 x 7-0	EA	82.00	280	362
1950	2-6 x 7-0	"	82.00	280	362
1970	3-0 x 7-0	"	82.00	270	352
1980	3-4 x 7-0	"	82.00	280	362
2000	Lauan faced				
2020	2-4 x 7-0	EA	82.00	190	272
2030	2-6 x 7-0	"	82.00	220	302
2060	3-4 x 7-0	"	82.00	240	322
2080	3-0 x 7-0	"	82.00	260	342
2140	Tempered hardboard faced				
2170	2-6 x 7-0	EA	82.00	280	362

DIVISION # 08 DOORS AND WINDOWS

08210.10	WOOD DOORS, Cont'd...	UNIT	LABOR	MAT.	TOTAL
2190	3-0 x 7-0	EA	82.00	320	402
2200	3-4 x 7-0	"	82.00	340	422
2250	Hollow core, 1-3/4" thick				
2270	Birch faced				
2295	2-6 x 7-0	EA	82.00	190	272
2320	3-0 x 7-0	"	82.00	200	282
2340	3-4 x 7-0	"	82.00	220	302
2400	Lauan faced				
2430	2-6 x 6-8	EA	82.00	130	212
2460	3-0 x 6-8	"	82.00	120	202
2480	3-4 x 6-8	"	82.00	120	202
2520	Tempered hardboard				
2550	2-6 x 7-0	EA	82.00	110	192
2580	3-0 x 7-0	"	82.00	120	202
2600	3-4 x 7-0	"	82.00	130	212
2620	Add-on, louver	"	65.00	35.00	100
2640	Glass	"	65.00	110	175
2700	Exterior doors, 3-0 x 7-0 x 2-1/2", solid core				
2710	Carved				
2720	One face	EA	160	1,460	1,620
2740	Two faces	"	160	2,020	2,180
3000	Closet doors, 1-3/4" thick				
3001	Bi-fold or bi-passing, includes frame and trim				
3020	Paneled				
3040	4-0 x 6-8	EA	110	510	620
3060	6-0 x 6-8	"	110	580	690
3070	Louvered				
3080	4-0 x 6-8	EA	110	350	460
3100	6-0 x 6-8	"	110	420	530
3130	Flush				
3140	4-0 x 6-8	EA	110	260	370
3160	6-0 x 6-8	"	110	330	440
3170	Primed				
3180	4-0 x 6-8	EA	110	280	390
3200	6-0 x 6-8	"	110	310	420
3210	French Door, Dual-Tempered, Clear-Glass, 6'-8'				
3220	24"x80"x1-3/4"	EA	130	460	590
3230	with Low-E glass	"	130	630	760
3240	30"x80"x1-3/4"	"	130	530	660
3250	with Low-E glass	"	130	630	760
3260	42"x80"x1-3/4"	"	130	690	820
3270	with Low-E glass	"	130	830	960
3280	24"x96"x1-3/4"	"	130	550	680
3290	with Low-E glass	"	130	690	820
3310	32"x96"x1-3/4"	"	130	620	750
3320	with Low-E glass	"	130	760	890
3330	French door, 10-lite, 1-3/4' thick, 6'-8" high				
3340	24" wide	EA	130	490	620
3350	30" wide	"	130	600	730
3360	36" wide	"	130	610	740
3370	96" high, 24" wide	"	130	620	750
3380	30" wide	"	130	750	880

DIVISION # 08 DOORS AND WINDOWS

		UNIT	LABOR	MAT.	TOTAL
08210.10	**WOOD DOORS, Cont'd...**				
3390	36" wide	EA	130	970	1,100
3400	French door, 1-lite, 1-3/4' thick, 6'-8" high				
3410	48" wide	EA	130	1,680	1,810
3420	56" wide	"	130	2,000	2,130
3430	60" wide	"	130	2,330	2,460
3440	72" wide	"	130	2,200	2,330
3500	Fiberglass, single door, 6'-8" high				
3510	2-4" wide	EA	82.00	270	352
3520	2-6" wide	"	82.00	400	482
3530	2-8" wide	"	82.00	400	482
3540	3-0" wide	"	82.00	490	572
3550	8'-0" high				
3560	2-4" wide	EA	82.00	530	612
3570	2-6" wide	"	82.00	670	752
3580	2-8" wide	"	82.00	800	882
3590	3-0" wide	"	82.00	760	842
3600	Fiberglass, double door, 6'-8" high				
3610	56" wide	EA	130	930	1,060
3620	60" wide	"	130	1,070	1,200
3630	64" wide	"	130	1,100	1,230
3640	72" wide	"	130	1,260	1,390
3650	8'-0" high				
3700	56" wide	EA	130	340	470
3710	60" wide	"	130	360	490
3720	64" wide	"	130	390	520
3730	72" wide	"	130	420	550
3740	Metal clad, double door, 6'-8" high				
3750	56" wide	EA	130	320	450
3760	60" wide	"	130	340	470
3770	64" wide	"	130	340	470
3780	72" wide	"	130	360	490
3790	8'-0" high				
3800	56" wide	EA	130	540	670
3810	60" wide	"	130	570	700
3820	64" wide	"	130	640	770
3830	72" wide	"	130	700	830
3840	For outswinging doors, add Minimum	"			80.00
3850	Maximum	"			170
08210.90	**WOOD FRAMES**				
0080	Frame, interior, pine				
0100	2-6 x 6-8	EA	93.00	100	193
0160	3-0 x 6-8	"	93.00	120	213
0180	5-0 x 6-8	"	93.00	120	213
0200	6-0 x 6-8	"	93.00	130	223
0220	2-6 x 7-0	"	93.00	120	213
0260	3-0 x 7-0	"	93.00	140	233
0280	5-0 x 7-0	"	130	150	280
0300	6-0 x 7-0	"	130	160	290
1000	Exterior, custom, with threshold, including trim				
1040	Walnut				
1060	3-0 x 7-0	EA	160	420	580

		UNIT	LABOR	MAT.	TOTAL
08210.90	**WOOD FRAMES, Cont'd...**				
1080	6-0 x 7-0	EA	160	480	640
1090	Oak				
1100	3-0 x 7-0	EA	160	380	540
1120	6-0 x 7-0	"	160	430	590
1200	Pine				
1240	2-6 x 7-0	EA	130	160	290
1300	3-0 x 7-0	"	130	180	310
1320	3-4 x 7-0	"	130	200	330
1340	6-0 x 7-0	"	220	210	430
3000	Fire-rated wood jambs				
3010	20-Minute, positive or neutral pressure, 3/4"	LF	5.43	10.75	16.18
3020	45-60-minute	"	5.43	12.00	17.43
3030	90-minute	"	5.43	13.25	18.68
3040	120-minute	"	5.43	16.00	21.43
08300.10	**SPECIAL DOORS**				
1000	Vault door and frame, class 5, steel	EA	650	7,940	8,590
1480	Overhead door, coiling insulated				
1500	Chain gear, no frame, 12' x 12'	EA	820	3,410	4,230
2000	Aluminum, bronze glass panels, 12-9 x 13-0	"	650	3,850	4,500
2200	Garage, flush, ins. metal, primed, 9-0 x 7-0	"	220	1,100	1,320
3000	Sliding fire doors, motorized, fusible link, 3 hr.				
3040	3-0 x 6-8	EA	1,300	5,620	6,920
3060	3-8 x 6-8	"	1,300	5,690	6,990
3080	4-0 x 8-0	"	1,300	5,790	7,090
3100	5-0 x 8-0	"	1,300	5,900	7,200
3200	Metal clad doors, including electric motor				
3210	Light duty				
3220	Minimum	SF	10.75	45.00	55.75
3240	Maximum	"	26.00	73.00	99.00
3250	Heavy duty				
3260	Minimum	SF	32.50	70.00	103
3280	Maximum	"	40.75	110	151
3500	Counter doors (roll-up shutters), std, manual				
3510	Opening, 4' high				
3520	4' wide	EA	540	1,300	1,840
3540	6' wide	"	540	1,760	2,300
3560	8' wide	"	590	1,980	2,570
3580	10' wide	"	820	2,200	3,020
3590	14' wide	"	820	2,750	3,570
3595	6' high				
3600	4' wide	EA	540	1,540	2,080
3620	6' wide	"	590	2,010	2,600
3630	8' wide	"	650	2,200	2,850
3640	10' wide	"	820	2,480	3,300
3650	14' wide	"	930	2,810	3,740
3660	For stainless steel, add to material, 40%				
3670	For motor operator, add	EA			1,520
3800	Service doors (roll-up shutters), std, manual				
3810	Opening				
3820	8' high x 8' wide	EA	360	1,650	2,010
3840	12' high x 12' wide	"	820	2,310	3,130

DIVISION # 08 DOORS AND WINDOWS

		UNIT	LABOR	MAT.	TOTAL
08300.10	**SPECIAL DOORS, Cont'd...**				
3860	16' high x 14' wide	EA	1,090	4,240	5,330
3870	20' high x 14' wide	"	1,630	4,790	6,420
3880	24' high x 16' wide	"	1,450	7,810	9,260
3890	For motor operator				
3900	Up to 12-0 x 12-0, add	EA			1,550
3920	Over 12-0 x 12-0, add	"			1,980
4040	Roll-up doors				
4050	13-0 high x 14-0 wide	EA	930	1,560	2,490
4060	12-0 high x 14-0 wide	"	930	1,980	2,910
5200	Sectional wood overhead, frames not incl.				
5220	Commercial grade, HD, 1-3/4" thick, manual				
5240	8' x 8'	EA	540	1,080	1,620
5260	10' x 10'	"	590	1,560	2,150
5280	12' x 12'	"	650	2,030	2,680
5290	Chain hoist				
5300	12' x 16' high	EA	1,090	3,000	4,090
5320	14' x 14' high	"	820	3,290	4,110
5340	20' x 8' high	"	1,300	2,830	4,130
5360	16' high	"	1,630	6,350	7,980
5990	Sectional metal overhead doors, complete				
6000	Residential grade, manual				
6020	9' x 7'	EA	260	740	1,000
6040	16' x 7'	"	330	1,330	1,660
6100	Commercial grade				
6120	8' x 8'	EA	540	860	1,400
6140	10' x 10'	"	590	1,150	1,740
6160	12' x 12'	"	650	1,910	2,560
6180	20' x 14', with chain hoist	"	1,300	4,560	5,860
6400	Sliding glass doors				
6420	Tempered plate glass, 1/4" thick				
6430	6' wide				
6440	Economy grade	EA	220	1,180	1,400
6450	Premium grade	"	220	1,350	1,570
6455	12' wide				
6460	Economy grade	EA	330	1,650	1,980
6465	Premium grade	"	330	2,480	2,810
6470	Insulating glass, 5/8" thick				
6475	6' wide				
6480	Economy grade	EA	220	1,450	1,670
6500	Premium grade	"	220	1,860	2,080
6505	12' wide				
6510	Economy grade	EA	330	1,800	2,130
6515	Premium grade	"	330	2,890	3,220
6520	1" thick				
6525	6' wide				
6530	Economy grade	EA	220	1,820	2,040
6540	Premium grade	"	220	2,100	2,320
6545	12' wide				
6550	Economy grade	EA	330	2,830	3,160
6560	Premium grade	"	330	4,140	4,470
6600	Added costs				
6620	Custom quality, add to material, 30%				

		UNIT	LABOR	MAT.	TOTAL
08300.10	**SPECIAL DOORS, Cont'd...**				
6630	Tempered glass, 6' wide, add	SF			5.08
6880	Residential storm door				
6900	Minimum	EA	110	180	290
6920	Average	"	110	240	350
6940	Maximum	"	160	530	690
08410.10	**STOREFRONTS**				
0135	Storefront, aluminum and glass				
0140	Minimum	SF	9.00	29.25	38.25
0150	Average	"	10.25	43.75	54.00
0160	Maximum	"	12.00	87.00	99.00
1020	Entrance doors, premium, closers, panic dev.,etc.				
1030	1/2" thick glass				
1040	3' x 7'	EA	600	3,890	4,490
1060	6' x 7'	"	900	6,650	7,550
1065	3/4" thick glass				
1070	3' x 7'	EA	600	4,030	4,630
1080	6' x 7'	"	900	6,720	7,620
1085	1" thick glass				
1090	3' x 7'	EA	600	4,370	4,970
1100	6' x 7'	"	900	7,720	8,620
1150	Revolving doors				
1151	7' diameter, 7' high				
1160	Minimum	EA	7,130	27,660	34,790
1170	Average	"	11,400	34,800	46,200
1180	Maximum	"	14,250	44,910	59,160
08510.10	**STEEL WINDOWS**				
0100	Steel windows, primed				
1000	Casements				
1010	Operable				
1020	Minimum	SF	4.23	52.00	56.23
1040	Maximum	"	4.80	78.00	82.80
1060	Fixed sash	"	3.60	41.25	44.85
1080	Double hung	"	4.00	78.00	82.00
1100	Industrial windows				
1120	Horizontally pivoted sash	SF	4.80	66.00	70.80
1130	Fixed sash	"	4.00	52.00	56.00
1135	Security sash				
1140	Operable	SF	4.80	82.00	86.80
1150	Fixed	"	4.00	73.00	77.00
1155	Picture window	"	4.00	35.25	39.25
1160	Projecting sash				
1170	Minimum	SF	4.50	61.00	65.50
1180	Maximum	"	4.50	75.00	79.50
1930	Mullions	LF	3.60	16.00	19.60
08520.10	**ALUMINUM WINDOWS**				
0110	Jalousie				
0120	3-0 x 4-0	EA	90.00	390	480
0140	3-0 x 5-0	"	90.00	450	540
0220	Fixed window				
0240	6 sf to 8 sf	SF	10.25	19.25	29.50
0250	12 sf to 16 sf	"	8.00	17.00	25.00

		UNIT	LABOR	MAT.	TOTAL
08520.10	**ALUMINUM WINDOWS, Cont'd...**				
0255	Projecting window				
0260	6 sf to 8 sf	SF	18.00	42.50	60.50
0270	12 sf to 16 sf	"	12.00	38.25	50.25
0275	Horizontal sliding				
0280	6 sf to 8 sf	SF	9.00	27.75	36.75
0290	12 sf to 16 sf	"	7.20	25.50	32.70
1140	Double hung				
1160	6 sf to 8 sf	SF	14.50	38.25	52.75
1180	10 sf to 12 sf	"	12.00	34.00	46.00
3010	Storm window, 0.5 cfm, up to				
3020	60 u.i. (united inches)	EA	36.00	89.00	125
3060	80 u.i.	"	36.00	100	136
3080	90 u.i.	"	40.00	100	140
3100	100 u.i.	"	40.00	110	150
3110	2.0 cfm, up to				
3120	60 u.i.	EA	36.00	110	146
3160	80 u.i.	"	36.00	120	156
3180	90 u.i.	"	40.00	130	170
3200	100 u.i.	"	40.00	130	170
08600.10	**WOOD WINDOWS**				
0980	Double hung				
0990	24" x 36"				
1000	Minimum	EA	65.00	240	305
1002	Average	"	82.00	350	432
1004	Maximum	"	110	470	580
1010	24" x 48"				
1020	Minimum	EA	65.00	280	345
1022	Average	"	82.00	410	492
1024	Maximum	"	110	570	680
1030	30" x 48"				
1040	Minimum	EA	72.00	290	362
1042	Average	"	93.00	410	503
1044	Maximum	"	130	590	720
1050	30" x 60"				
1060	Minimum	EA	72.00	320	392
1062	Average	"	93.00	510	603
1064	Maximum	"	130	630	760
1160	Casement				
1180	1 leaf, 22" x 38" high				
1220	Minimum	EA	65.00	350	415
1222	Average	"	82.00	430	512
1224	Maximum	"	110	500	610
1230	2 leaf, 50" x 50" high				
1240	Minimum	EA	82.00	940	1,022
1242	Average	"	110	1,230	1,340
1244	Maximum	"	160	1,410	1,570
1250	3 leaf, 71" x 62" high				
1260	Minimum	EA	82.00	1,550	1,632
1262	Average	"	110	1,580	1,690
1264	Maximum	"	160	1,890	2,050
1290	5 leaf, 119" x 75" high				

		UNIT	LABOR	MAT.	TOTAL
08600.10	**WOOD WINDOWS, Cont'd...**				
1300	Minimum	EA	93.00	2,670	2,763
1302	Average	"	130	2,880	3,010
1304	Maximum	"	220	3,680	3,900
1360	Picture window, fixed glass, 54" x 54" high				
1400	Minimum	EA	82.00	550	632
1422	Average	"	93.00	620	713
1424	Maximum	"	110	1,100	1,210
1430	68" x 55" high				
1440	Minimum	EA	82.00	990	1,072
1442	Average	"	93.00	1,140	1,233
1444	Maximum	"	110	1,490	1,600
1480	Sliding, 40" x 31" high				
1520	Minimum	EA	65.00	330	395
1522	Average	"	82.00	500	582
1524	Maximum	"	110	600	710
1530	52" x 39" high				
1540	Minimum	EA	82.00	410	492
1542	Average	"	93.00	610	703
1544	Maximum	"	110	650	760
1550	64" x 72" high				
1560	Minimum	EA	82.00	630	712
1562	Average	"	110	1,010	1,120
1564	Maximum	"	130	1,110	1,240
1760	Awning windows				
1780	34" x 21" high				
1800	Minimum	EA	65.00	330	395
1822	Average	"	82.00	380	462
1824	Maximum	"	110	440	550
1880	48" x 27" high				
1900	Minimum	EA	72.00	410	482
1902	Average	"	93.00	490	583
1904	Maximum	"	130	570	700
1920	60" x 36" high				
1940	Minimum	EA	82.00	430	512
1942	Average	"	110	760	870
1944	Maximum	"	130	860	990
8000	Window frame, milled				
8010	Minimum	LF	13.00	6.09	19.09
8020	Average	"	16.25	6.80	23.05
8030	Maximum	"	21.75	10.25	32.00
08710.10	**HINGES**				
1200	Hinges, material only				
1250	3 x 3 butts, steel, interior, plain bearing	PAIR			20.75
1260	4 x 4 butts, steel, standard	"			30.50
1270	5 x 4-1/2 butts, bronze/s. steel, heavy duty	"			79.00
1290	Pivot hinges				
1300	Top pivot	EA			88.00
1310	Intermediate pivot	"			94.00
1320	Bottom pivot	"			180

		UNIT	LABOR	MAT.	TOTAL
08710.20	**LOCKSETS**				
1280	Latchset, heavy duty				
1300	Cylindrical	EA	40.75	190	231
1320	Mortise	"	65.00	200	265
1325	Lockset, heavy duty				
1330	Cylindrical	EA	40.75	310	351
1350	Mortise	"	65.00	350	415
2200	Preassembled locks and latches, brass				
2220	Latchset, passage or closet latch	EA	54.00	280	334
2225	Lockset				
2230	Privacy (bath or bathroom)	EA	54.00	340	394
2240	Entry lock	"	54.00	490	544
2285	Lockset				
2290	Privacy (bath or bedroom)	EA	54.00	220	274
2300	Entry lock	"	54.00	240	294
08710.30	**CLOSERS**				
2600	Door closers				
2610	Standard	EA	82.00	240	322
2620	Heavy duty	"	82.00	280	362
08710.40	**DOOR TRIM**				
1600	Panic device				
1610	Mortise	EA	160	810	970
1620	Vertical rod	"	160	1,230	1,390
1630	Labeled, rim type	"	160	850	1,010
1640	Mortise	"	160	1,110	1,270
1650	Vertical rod	"	160	1,180	1,340
2300	Door plates				
2305	Kick plate, aluminum, 3 beveled edges				
2310	10" x 28"	EA	32.50	30.50	63.00
2340	10" x 38"	"	32.50	39.75	72.25
2350	Push plate, 4" x 16"				
2360	Aluminum	EA	13.00	27.75	40.75
2371	Bronze	"	13.00	88.00	101
2380	Stainless steel	"	13.00	70.00	83.00
2385	Armor plate, 40" x 34"	"	26.00	81.00	107
2388	Pull handle, 4" x 16"				
2390	Aluminum	EA	13.00	98.00	111
2400	Bronze	"	13.00	190	203
2420	Stainless steel	"	13.00	140	153
2425	Hasp assembly				
2430	3"	EA	10.75	4.62	15.37
2440	4-1/2"	"	14.50	5.77	20.27
2450	6"	"	18.75	9.18	27.93
08710.60	**WEATHERSTRIPPING**				
0100	Weatherstrip, head and jamb, metal strip, neoprene bulb				
0140	Standard duty	LF	3.62	5.19	8.81
0160	Heavy duty	"	4.07	5.77	9.84
3980	Spring type				
4000	Metal doors	EA	160	55.00	215
4010	Wood doors	"	220	55.00	275
4020	Sponge type with adhesive backing	"	65.00	51.00	116
4500	Thresholds				

DIVISION # 08 DOORS AND WINDOWS

		UNIT	LABOR	MAT.	TOTAL
08710.60	**WEATHERSTRIPPING, Cont'd...**				
4510	Bronze	LF	16.25	53.00	69.25
4515	Aluminum				
4520	Plain	LF	16.25	38.50	54.75
4525	Vinyl insert	"	16.25	39.25	55.50
4530	Aluminum with grit	"	16.25	37.50	53.75
4533	Steel				
4535	Plain	LF	16.25	29.75	46.00
4540	Interlocking	"	54.00	39.50	93.50
08810.10	**GLASS GLAZING**				
0800	Sheet glass, 1/8" thick	SF	4.00	8.91	12.91
1020	Plate glass, bronze or grey, 1/4" thick	"	6.54	13.00	19.54
1040	Clear	"	6.54	10.25	16.79
1060	Polished	"	6.54	12.00	18.54
1980	Plexiglass				
2000	1/8" thick	SF	6.54	5.73	12.27
2020	1/4" thick	"	4.00	10.25	14.25
3000	Float glass, clear				
3010	3/16" thick	SF	6.00	6.93	12.93
3020	1/4" thick	"	6.54	7.07	13.61
3040	3/8" thick	"	9.00	14.25	23.25
3100	Tinted glass, polished plate, twin ground				
3120	3/16" thick	SF	6.00	9.55	15.55
3130	1/4" thick	"	6.54	9.55	16.09
3140	3/8" thick	"	9.00	15.25	24.25
5000	Insulated glass, bronze or gray				
5020	1/2" thick	SF	12.00	19.50	31.50
5040	1" thick	"	18.00	23.25	41.25
5100	Spandrel, polished, 1 side, 1/4" thick	"	6.54	15.50	22.04
6000	Tempered glass (safety)				
6010	Clear sheet glass				
6020	1/8" thick	SF	4.00	10.75	14.75
6030	3/16" thick	"	5.53	13.00	18.53
6040	Clear float glass				
6050	1/4" thick	SF	6.00	11.25	17.25
6060	5/16" thick	"	7.20	20.00	27.20
6070	3/8" thick	"	9.00	24.50	33.50
6080	1/2" thick	"	12.00	33.50	45.50
6800	Insulating glass, two lites, clear float glass				
6840	1/2" thick	SF	12.00	13.75	25.75
6850	5/8" thick	"	14.50	16.00	30.50
6860	3/4" thick	"	18.00	17.50	35.50
6870	7/8" thick	"	20.50	18.50	39.00
6880	1" thick	"	24.00	24.75	48.75
6885	Glass seal edge				
6890	3/8" thick	SF	12.00	11.75	23.75
6895	Tinted glass				
6900	1/2" thick	SF	12.00	23.75	35.75
6910	1" thick	"	24.00	25.50	49.50
6920	Tempered, clear				
6930	1" thick	SF	24.00	46.50	70.50
7200	Plate mirror glass				

08 - 12

		UNIT	LABOR	MAT.	TOTAL
08810.10	**GLASS GLAZING, Cont'd...**				
7205	1/4" thick				
7210	15 sf	SF	7.20	11.75	18.95
7220	Over 15 sf	"	6.54	10.75	17.29
9650	Sand-Blasted Glass				
9660	3/16" Float glass, full sandblast, no custom decoration	SF	6.00	12.25	18.25
9670	3/8" Float glass, full sandblast, no custom decoration	"	7.20	14.50	21.70
9680	Beveled glass				
9690	commercial standard grade	SF	7.20	150	157
08910.10	**GLAZED CURTAIN WALLS**				
1000	Curtain wall, aluminum system, framing sections				
1005	2" x 3"				
1010	Jamb	LF	6.00	19.50	25.50
1020	Horizontal	"	6.00	19.75	25.75
1030	Mullion	"	6.00	26.50	32.50
1035	2" x 4"				
1040	Jamb	LF	9.00	26.50	35.50
1060	Horizontal	"	9.00	27.25	36.25
1070	Mullion	"	9.00	26.50	35.50
1080	3" x 5-1/2"				
1090	Jamb	LF	9.00	35.00	44.00
1100	Horizontal	"	9.00	38.75	47.75
1110	Mullion	"	9.00	35.25	44.25
1115	4" corner mullion	"	12.00	46.75	58.75
1120	Coping sections				
1130	1/8" x 8"	LF	12.00	44.25	56.25
1140	1/8" x 9"	"	12.00	44.50	56.50
1150	1/8" x 12-1/2"	"	14.50	45.75	60.25
1160	Sill section				
1170	1/8" x 6"	LF	7.20	43.75	50.95
1180	1/8" x 7"	"	7.20	44.25	51.45
1190	1/8" x 8-1/2"	"	7.20	45.00	52.20
1200	Column covers, aluminum				
1210	1/8" x 26"	LF	18.00	43.75	61.75
1220	1/8" x 34"	"	19.00	48.50	67.50
1230	1/8" x 38"	"	19.00	49.00	68.00
1500	Doors				
1600	Aluminum framed, standard hardware				
1620	Narrow stile				
1630	2-6 x 7-0	EA	360	810	1,170
1640	3-0 x 7-0	"	360	810	1,170
1660	3-6 x 7-0	"	360	840	1,200
1700	Wide stile				
1720	2-6 x 7-0	EA	360	1,380	1,740
1730	3-0 x 7-0	"	360	1,490	1,850
1750	3-6 x 7-0	"	360	1,600	1,960
1800	Flush panel doors, to match adjacent wall panels				
1810	2-6 x 7-0	EA	450	1,170	1,620
1820	3-0 x 7-0	"	450	1,230	1,680
1840	3-6 x 7-0	"	450	1,270	1,720
2100	Wall panel, insulated				
2120	U=.08	SF	6.00	14.75	20.75

		UNIT	LABOR	MAT.	TOTAL
08910.10	**GLAZED CURTAIN WALLS, Cont'd...**				
2140	U=.10	SF	6.00	14.00	20.00
2160	U=.15	"	6.00	12.50	18.50
3000	Window wall system, complete				
3010	Minimum	SF	7.20	42.75	49.95
3030	Average	"	8.00	68.00	76.00
3050	Maximum	"	10.25	160	170
4860	Added costs				
4870	For bronze, add 20% to material				
4880	For stainless steel, add 50% to material				

		UNIT	COST
08999.10	**DOORS**		
0900	HOLLOW METAL		
1000	CUSTOM FRAMES (16 ga)		
1010	2'6" x 6'8" x 4-3/4"	EA	370
1020	2'8" x 6'8" x 4-3/4"	"	410
1030	3'0" x 6'8" x 4-3/4"	"	430
1040	3'4" x 6'8" x 4-3/4"	"	450
2000	Add for A, B or C Label	"	48.25
2010	Add for Side Lights	"	221
2020	Add for Frames over 7'0" - Height	"	89.50
2030	Add for Frames over 6-3/4" - Width	"	77.75
3000	CUSTOM DOORS (1-3/8" or 1-3/4") (18 ga)		
3010	2'6" x 6'8" x 1-3/4"	EA	550
3020	2'8" x 6'8" x 1-3/4"	"	570
3030	3'0" x 6'8" x 1-3/4"	"	590
3040	3'4" x 6'8" x 1-3/4"	"	600
4000	Add for 16 ga	"	86.50
4010	Add for B and C Label	"	38.75
4020	Add for Vision Panels or Lights	"	166
5000	STOCK FRAMES (16 ga)		
5010	2'6" x 6'8" or 7'0" x 4-3/4"	EA	251
5020	2'8" x 6'8" or 7'0" x 4-3/4"	"	265
5030	3'0" x 6'8" or 7'0" x 4-3/4"	"	275
5040	3'4" x 6'8" or 7'0" x 4-3/4"	"	295
5050	Add for A, B or C Label	"	45.75
6000	STOCK DOORS (1-3/8" or 1-3/4" - 18 ga)		
6010	2'6" x 6'8" or 7'0"	EA	580
6020	2'8" x 6'8" or 7'0"	"	600
6030	3'0" x 6'8" or 7'0"	"	600
6040	3'4" x 6'8" or 7'0"	"	660
6050	Add for B or C Label	"	36.00
08999.20	**WOOD DOORS (Incl. butts and locksets)**		
1000	FLUSH DOORS - No Label - Paint Grade		
2000	Birch - 7 Ply 2'4" x 6'8" x 1-3/8" -Hollow Core	EA	231
2010	2'6" x 6'8" x 1-3/8"	"	241
2020	2'8" x 6'8" x 1-3/8"	"	255
2030	3'0" x 6'8" x 1-3/8"	"	255
2040	2'4" x 6'8" x 1-3/4"	"	255
2050	2'6" x 6'8" x 1-3/4"	"	255
2060	2'8" x 6'8" x 1-3/4"	"	265
2070	3'0" x 6'8" x 1-3/4"	"	275
2080	3'4" x 6'8" x 1-3/4"	"	275
2090	3'6" x 6'8" x 1-3/4"	"	275
3000	Birch - 7 Ply 2'4" x 6'8" x 1-3/8" -Solid Core	"	275
3010	2'6" x 6'8" x 1-3/8"	"	275
3020	2'8" x 6'8" x 1-3/8"	"	315
3030	3'0" x 6'8" x 1-3/8"	"	315
3040	2'4" x 6'8" x 1-3/4"	"	315
3050	2'6" x 6'8" x 1-3/4"	"	315
3060	2'8" x 6'8" x 1-3/4"	"	340
3070	3'0" x 6'8" x 1-3/4"	"	340
3080	3'4" x 6'8" x 1-3/4"	"	350

DIVISION # 08 DOORS AND WINDOWS - QUICK ESTIMATING

		UNIT	COST
08999.20	**WOOD DOORS (Incl. butts and locksets), Cont'd...**		
3090	3'6" x 6'8" x 1-3/4"E	EA	410
4000	Add for Stain Grade	"	45.00
4010	Add for 5 Ply	"	97.75
4020	Add for Jamb & Trim - Solid - Knock Down	"	101
4030	Add for Jamb & Trim - Veneer - Knock Down	"	66.50
4040	Add for Red Oak - Rotary Cut	"	40.25
4050	Add for Red Oak - Plain Sliced	"	54.00
5000	Deduct for Lauan	"	42.25
5010	Add for Architectural Grade - 7 Ply	"	94.50
5020	Add for Vinyl Overlay	"	73.00
5030	Add for 7' - 0" Doors	"	32.00
5040	Add for Lite Cutouts with Metal Frame	"	106
5050	Add for Wood Louvers	"	167
5060	Add for Transom Panels & Side Panels	"	211
6000	LABELED - 1-3/4" - Paint Grade		
6010	2'6" x 6'8" - Birch - 20 minimum	EA	370
6020	2'8" x 6'8"	"	350
6030	3'0" x 6'8"	"	360
6040	3'4" x 6'8"	"	410
6050	3'6" x 6'8"	"	470
7000	2'6" x 6'8" - Birch - 45 Min	"	390
7010	2'8" x 6'8"	"	410
7020	3'0" x 6'8"	"	420
7030	3'4" x 6'8"	"	440
7040	3'6" x 6'8"	"	460
8000	2'6" x 6'8" - Birch - 60 Min	"	510
8010	2'8" x 6'8"	"	550
8020	3'0" x 6'8"	"	560
8030	3'4" x 6'8"	"	590
8040	3'6" x 6'8"	"	460
9000	PANEL DOORS		
9010	Exterior - 1-3/4" -Pine		
9020	2'8" x 6'8"	EA	810
9030	3'0" x 6'8"	"	780
9040	Interior - 1-3/8" -Pine		
9050	2'6" x 6'8"	EA	420
9060	2'8" x 6'8"	"	460
9070	3'0" x 6'8"	"	490
9080	Add for Sidelights	"	360
9140	Exterior - 1-3/4"- Birch/ Oak		
9150	2'8" x 6'8"	EA	860
9160	3'0" x 6'8"	"	850
9170	Interior - 1-3/8"		
9180	2'6" x 6'8"	EA	590
9190	2'8" x 6'8"	"	610
9200	3'0" x 6'8"	"	650
9210	Add for Sidelights	"	570
9212	LOUVERED DOORS - 1-3/8" -Pine		
9213	2'0" x 6'8"	EA	320
9214	2'6" x 6'8"	"	370
9215	2'8" x 6'8"	"	390
9216	3'0" x 6'8"	"	410

		UNIT	COST
08999.20	**WOOD DOORS (Incl. butts and locksets), Cont'd...**		
9220	Birch/Oak - 1-3/8"		
9230	2'0" x 6'8"	EA	680
9240	2'6" x 6'8"	"	690
9250	2'8" x 6'8"	"	720
9260	3'0" x 6'8"	"	730
9270	BI-FOLD with hdwr. 1-3/8" - Flush		
9280	Birch/Oak, 2'0" x 6'8"	EA	235
9290	2'4" x 6'8"	"	260
9300	2'6" x 6'8"	"	270
9320	Lauan, 2'0" x 6'8"	"	235
9330	2'4" x 6'8"	"	245
9340	2'6" x 6'8"	"	255
9350	Add for Prefinished	"	29.25
9360	CAFE DOORS - 1-1/8" - Pair		
9370	2'6" x 3'8"	EA	470
9380	2'8" x 3'8"	"	490
9390	3'0" x 3'8"	"	500
9400	FRENCH DOORS - 1-3/8" -Pine		
9410	2'6" x 3'8"	EA	630
9420	2'8" x 3'8"	"	650
9430	3'0" x 3'8"	"	680
9440	2'6" x 3'8" -Birch/Oak	"	810
9450	2'8" x 3'8"	"	820
9460	3'0" x 3'8"	"	830
9470	DUTCH DOORS		
9480	2'6" x 3'8"	EA	940
9490	2'8" x 3'8"	"	1,030
9500	3'0" x 3'8"	"	1,310
9510	PREHUNG DOORS (INCL. TRIM)		
9520	Exterior, entrance - 1-3/4"		
9530	Panel - 2'8" x 6'8"	EA	830
9540	3'0" x 6'8"	"	910
9550	3'0" x 7'0"	"	1,020
9560	Add for Insulation	"	285
9570	Add for Side Lights	"	345
9580	Interior - 1-3/8"		
9590	Flush, H.C. - 2'6" x 6'8" - Birch/Oak	EA	450
9600	2'8" x 6'8"	"	460
9610	3'0" x 6'8"Each	"	480
9620	Flush, H.C. - 2'6" x 6'8" - Lauan	"	350
9630	2'8" x 6'8"	"	390
9640	3'0" x 6'8"Each	"	390
9650	Add for Int. Panel Door	"	320
08999.30	**SPECIAL DOORS**		
1000	BI-FOLDING - PREHUNG, 1-3/8"		
1020	Wood - 2 Door - 2'0" x 6'8" - Oak/flush	EA	370
1030	2'6" x 6'8"	"	410
1040	3'0" x 6'8"	"	430
1050	4 Door - 4'0" x 6'8"	"	470
1060	5'0" x 6'8"	"	480
1070	6'0" x 6'8"	"	570

		UNIT	COST
08999.30	**SPECIAL DOORS, Cont'd...**		
1080	Add for Prefinished	EA	44.25
1090	Add for Jambs and Casings	"	109
2000	Wood - 2 Door - 2'0" x 6'8" Pine/Panel	"	720
2010	2'6" x 6'8"	"	740
2020	3'0" x 6'8"	"	860
2030	4 Door - 4'0" x 6'8"	"	1,070
2040	5'0" x 6'8"	"	1,170
2050	6'0" x 6'8"	"	1,430
3000	Metal - 2 Door - 2'0" x 6'8"	"	221
3010	2'6" x 6'8"	"	231
3020	3'0" x 6'8"	"	285
3030	4 Door - 4'0" x 6'8"	"	305
3040	5'0" x 6'8"	"	330
3050	6'0" x 6'8"	"	360
3060	Add for Plastic Overlay	"	48.50
3070	Add for Louver or Decorative Type	"	250
4000	Leaded Mirror (Based on 2-Panel Unit)		
4010	4'0" x 6'8"	EA	1,040
4020	5'0" x 6'8"	"	1,220
4030	6'0" x 6'8"	"	1,220
7000	ROLLING - DOORS & GRILLES		
7010	Doors - 8' x 8'	EA	4,080
7020	10' x 10'	"	4,260
7030	Grilles - 6'-8" x 3'-2"	"	1,420
7040	6'-8" x 4'-2"	"	1,600
8000	SHOWER DOORS - 28" x 66"	"	420
9000	SLIDING OR PATIO DOORS		
9010	Metal (Aluminum) - Including Glass Thresholds & Screen		
9020	Opening 8'0" x 6'8"	EA	2,400
9030	8'0" x 8'0"	"	2,730
9040	Wood		
9050	Vinyl Clad 6'0" x 6'10"	EA	3,000
9060	8'0" x 6'10"	"	3,280
9070	Pine - Prefinished 6'0" x 6'10"	"	2,980
9080	8'0" x 6'10"	"	3,300
9090	Add for Grilles	"	480
9100	Add for Triple Glazing	"	315
9110	Add for Screen	"	221
9120	SOUND REDUCTION - Metal	"	2,890
9130	Wood	"	1,900
9200	VAULT DOORS - 6'6" x 2'2" with Frame - 2 hour	"	5,840
08999.50	**METAL WINDOWS**		
1000	EACH, 2'4" x 4'6" - ALUMINUM WINDOWS		
1010	Casement and Awning	EA	531
1020	Sliding or Horizontal	"	441
1030	Double and Single-Hung	"	471
1040	or Vertical Sliding		
1050	Projected	EA	501
1060	Add for Screens	"	75.50
1070	Add for Storms	"	132
1080	Add for Insulated Glass	"	115

		UNIT	COST
08999.50	**METAL WINDOWS, Cont'd...**		
2000	ALUMINUM SASH		
2010	Casement	EA	461
2020	Sliding	"	411
2030	Single-Hung	"	411
2040	Projected	"	337
2050	Fixed	"	347
3000	STEEL WINDOWS		
3010	Double-Hung	EA	890
3020	Projected	"	850
4000	STEEL SASH		
4010	Casement	EA	655
4020	Double-Hung	"	725
4030	Projected	"	561
4040	Fixed	"	571
5000	By S.F., 2'4" x 4'6" - ALUMINUM WINDOWS		
5010	Casement and Awning	SF	51.14
5020	Sliding or Horizontal	"	40.33
5030	Double and Single-Hung	"	45.12
5040	or Vertical Sliding	"	36.25
5050	Projected	"	50.91
5060	Add for Screens	"	7.24
5070	Add for Storms	"	13.32
5080	Add for Insulated Glass	"	10.98
6000	ALUMINUM SASH		
6010	Casement	SF	47.75
6020	Sliding	"	40.33
6030	Single-Hung	"	40.33
6040	Projected	"	34.37
6050	Fixed	"	35.58
7000	STEEL WINDOWS		
7010	Double-Hung	SF	79.50
7020	Projected	"	76.00
8000	STEEL SASH		
8010	Casement	SF	61.90
8020	Double-Hung	"	76.00
8030	Projected	"	52.60
8040	Fixed	"	54.37
08999.60	**WOOD WINDOWS**		
1000	BASEMENT OR UTILITY		
1020	Prefinished with Screen, 2'8" x 1'4"	EA	277
1030	2'8" x 2'0"	"	331
1040	Add for Double Glazing or Storm	"	70.75
2000	CASEMENT OR AWNING		
2010	Operating Units - Insulating Glass		
2020	Single 2'4" x 4'0"	EA	780
2030	2'4" x 5'0"	"	740
3000	Double with Screen 4'0" x 4'0" (2 units)	"	1,000
3010	4'0" x 5'0" (2 units)	"	1,290
3020	Triple with Screen 6'0" x 4'0" (3 units)	"	1,730
3030	6'0" x 5'0" (3 units)	"	1,750
4000	Fixed Units - Insulating Glass		

DIVISION # 08 DOORS AND WINDOWS - QUICK ESTIMATING

		UNIT	COST
08999.60	**WOOD WINDOWS, Cont'd...**		
4010	Single 2'4" x 4'0"	EA	640
4020	2'4" x 5'0"	"	720
4030	Picture 4'0" x 4'6"	"	890
4040	4'0" x 6'0"	"	1,050
5000	DOUBLE HUNG - Insul. Glass - 2'6" x 3'6"	"	690
5010	2'6" x 4'2"	"	690
5020	3'2" x 3'6"	"	710
5030	3'2" x 4'2"	"	780
5040	Add for Triple Glazing	"	110
6000	GLIDER - & Insul. Glass with Screen - 4'0" x 3'6"	"	1,610
6010	5'0" x 4'0"	"	1,790
8000	PICTURE WINDOWS		
8010	9'6" x 4'10"	EA	3,360
8020	9'6" x 5'6"	"	2,580
9000	CASEMENT - Insulating Glass		
9010	30° - 5'10" x 4'2" - (3 units)	EA	2,450
9015	7'10" x 4'2" - (4 units)	"	2,950
9020	5'10" x 5'2" - (3 units)	"	2,560
9030	7'10" x 5'2" - (4 units)	"	2,890
9040	45° - 5'4" x 4'2" - (3 units)	"	2,440
9050	7'4" x 4'2" - (4 units)	"	2,670
9060	5'4" x 5'2" - (3 units)	"	2,830
9070	7'4" x 5'2" - (4 units)	"	3,180
9100	CASEMENT BOW WINDOWS - Insulating Glass		
9110	6'2" x 4'2" - (3 units)	EA	2,550
9120	8'2" x 4'2" - (4 units)	"	3,360
9130	6'2" x 5'2" - (3 units)	"	2,840
9140	8'2" x 5'2" - (4 units)	"	3,690
9150	Add for Triple Glazing - per unit	"	95.75
9160	Add for Bronze Glazing - per unit	"	95.75
9170	Deduct for Primed Only - per unit	"	24.04
9180	Add for Screens - per unit	"	29.44
9200	90° BOX BAY WINDOWS - Insulating Glass 4'8" x 4'2"	"	3,360
9210	6'8" x 4'2"	"	4,250
9220	6'8" x 5'2"	"	4,410
9300	ROOF WINDOWS 1'10" x 3'10"-Fixed	"	980
9310	2'4" x 3'10"	"	1,120
9320	3'8" x 3'10"	"	1,390
9400	ROOF WINDOWS 1'10" x 3'10"-Movable	"	1,320
9410	2'4" x 3'10"	"	1,620
9420	3'8" x 3'10"	"	1,860
9500	CIRCULAR TOPS & ROUNDS - 4'0"	"	1,050
9510	6'0"	"	2,350
08999.70	**SPECIAL WINDOWS**		
1000	LIGHT-PROOF WINDOWS	SF	60.25
2000	PASS WINDOWS	"	46.50
3000	DETENTION WINDOWS	"	62.75
4000	VENETIAN BLIND WINDOWS (ALUMINUM)	"	52.75
5000	SOUND-CONTROL WINDOWS	"	51.50

08A - 20

		UNIT	COST
08999.80	**DOOR AND WINDOW ACCESSORIES**		
1000	STORMS AND SCREENS		
1010	Windows		
1020	Screen Only - Wood - 3' x 5'	EA	156
1030	Aluminum - 3' x 5'	"	181
1040	Storm & Screen Combination - Aluminum	"	204
1050	Doors	"	
1060	Screen Only - Wood	"	371
1070	Aluminum - 3' x 6' - 8'	"	371
1080	Storm & Screen Combination - Aluminum	"	445
1090	Wood 1-1/8"	"	490
2000	DETENTION SCREENS		
2010	Example: 4' 0" x 7' 0"	EA	1,100
3000	DOOR OPENING ASSEMBLIES		
3010	Floor or Overhead Electric Eye Units		
3020	Swing - Single 3' x 7' Door - Hydraulic	EA	6,630
3030	Double 6' x 7' Door - Hydraulic	"	10,560
3040	Sliding - Single 3' x 7' Door - Hydraulic	"	8,080
3050	Double 5' x 7' Doors- Hydraulic	"	10,510
3060	Industrial Doors - 10' x 8'	"	11,800
4000	SHUTTERS		
4010	16" x 1-1/8" x 48"	EA	178
4020	16" x 1-1/8" x 60"	"	214
4030	16" x 1-1/8" x 72"	"	254
08999.90	**FINISH HARDWARE**		
1000	BUTT HINGES -Painted		
1010	3" x 3"	EA	37.00
1020	3-1/2" x 3-1/2"	"	38.75
1030	4" x 4"	"	41.50
1040	4-1/2" x 4-1/2"	"	45.25
1050	4" x 4" Ball Bearing	"	68.00
1060	4-1/2" x 4-1/2" Ball Bearing	"	73.00
2000	Bronze		
2010	3" x 3"	EA	38.75
2020	3-1/2" x 3-1/2"	"	39.75
2030	4" x 4"	"	41.50
2040	4-1/2" x 4-1/2"	"	48.00
2050	4" x 4" Ball Bearing	"	70.00
2060	4-1/2" x 4-1/2" Ball Bearing	"	79.00
3000	Chrome		
3010	3" x 3"	EA	41.50
3020	3-1/2" x 3-1/2"	"	44.25
3030	4" x 4"	"	46.25
3040	4-1/2" x 4-1/2"	"	55.00
3050	4" x 4" Ball Bearing	"	79.00
3060	4-1/2" x 4-1/2" Ball Bearing	"	84.00
5000	CLOSERS - SURFACE MOUNTED - 3' - 0" Door	"	290
5010	3' - 4"	"	310
5020	3' - 8"	"	310
5030	4' - 0"	"	390
6000	CLOSERS - CONCEALED - Interior	"	530
6010	ExteriorEach	"	720

		UNIT	COST
08999.90	**FINISH HARDWARE, Cont'd...**		
6020	Add for Fusible Link - Electric	EA	290
6030	CLOSERS - FLOOR HINGES - Interior		740
6040	Exterior	"	1,140
6050	Add for Hold Open Feature		117
6060	Add for Double Acting Feature	"	380
7000	DEAD BOLT LOCK - Cylinder - Outside Key	"	250
7010	Cylinder - Double Key		360
7020	Flush - Push/ Pull		96.75
8000	EXIT DEVICES (PANIC) - Surface	"	1,050
8010	Mortise Lock	"	1,370
8020	Concealed		3,400
8030	Handicap (ADA) Automatic		2,480
9000	HINGES, SPRING (PAINTED) - 6" Single Acting	"	183
9010	6" Double Acting	"	221
9101	LATCHSETS -Bronze or Chrome	"	290
9102	Stainless Steel	"	330
9200	LOCKSETS - Mortise - Bronze or Chrome - H.D.	"	400
9210	Mortise - Bronze or Chrome - S.D.	"	320
9220	Stainless Steel	"	520
9230	Cylindrical - Bronze or Chrome	"	350
9240	Stainless Steel	"	410
9300	LEVER HANDICAP - Latch Set	"	410
9310	Lock Set	"	480
9400	PLATES - Kick - 8" x 34" - Aluminum	"	129
9410	Bronze	"	119
9420	Push - 6" x 15" - Aluminum	"	83.50
9430	Bronze	"	121
9440	Push & Pull Combination - Aluminum	"	140
9450	Bronze	"	201
9500	STOPS AND HOLDERS		
9510	Holder - Magnetic (No Electric)	EA	280
9520	Bumper	"	67.25
9530	Overhead - Bronze, Chrome or Aluminum	"	155
9540	Wall Stops	"	58.50
9550	Floor Stops	"	73.50
08999.91	**WEATHERSTRIPPING**		
1000	ASTRAGALS - Aluminum - 1/8" x 2"	EA	79.00
1010	Painted Steel	"	71.75
2000	DOORS (WOOD) - Interlocking	"	127
2010	Spring Bronze	"	138
2020	Add for Metal Doors	"	63.50
3000	SWEEPS - 36" WOOD DOORS - Aluminum	"	37.25
3010	Vinyl	"	36.25
4000	THRESHOLDS - Aluminum - 4" x 1/2"	"	75.25
4010	Bronze 4" x 1/2"	"	96.00
4020	5-1/2" x 1/2"	"	102
5000	WINDOWS (WOOD) - Interlocking	"	109
5010	Spring Bronze	"	127

TABLE OF CONTENTS   PAGE

09110.10	METAL STUDS	UNIT	LABOR	MAT.	TOTAL
0060	Studs, non load bearing, galvanized				
0130	3-5/8", 20 ga.				
0140	12" o.c.	SF	1.63	0.81	2.44
0142	16" o.c.	"	1.30	0.62	1.92
0144	24" o.c.	"	1.08	0.47	1.55
0170	25 ga.				
0180	12" o.c.	SF	1.63	0.53	2.16
0182	16" o.c.	"	1.30	0.44	1.74
0184	24" o.c.	"	1.08	0.33	1.41
0210	6", 20 ga.				
0220	12" o.c.	SF	2.03	1.14	3.17
0222	16" o.c.	"	1.63	0.83	2.46
0224	24" o.c.	"	1.35	0.68	2.03
0230	25 ga.				
0240	12" o.c.	SF	2.03	0.73	2.76
0242	16" o.c.	"	1.63	0.58	2.21
0244	24" o.c.	"	1.35	0.44	1.79
0980	Load bearing studs, galvanized				
0990	3-5/8", 16 ga.				
1000	12" o.c.	SF	1.63	1.47	3.10
1020	16" o.c.	"	1.30	1.36	2.66
1110	18 ga.				
1130	12" o.c.	SF	1.08	1.15	2.23
1140	16" o.c.	"	1.30	1.05	2.35
1980	6", 16 ga.				
2000	12" o.c.	SF	2.03	1.98	4.01
2001	16" o.c.	"	1.63	1.78	3.41
3000	Furring				
3160	On beams and columns				
3170	7/8" channel	LF	4.34	0.52	4.86
3180	1-1/2" channel	"	5.01	0.62	5.63
4460	On ceilings				
4470	3/4" furring channels				
4480	12" o.c.	SF	2.71	0.37	3.08
4490	16" o.c.	"	2.60	0.29	2.89
4495	24" o.c.	"	2.33	0.20	2.53
4500	1-1/2" furring channels				
4520	12" o.c.	SF	2.96	0.62	3.58
4540	16" o.c.	"	2.71	0.47	3.18
4560	24" o.c.	"	2.50	0.31	2.81
5000	On walls				
5020	3/4" furring channels				
5050	12" o.c.	SF	2.17	0.37	2.54
5100	16" o.c.	"	2.03	0.29	2.32
5150	24" o.c.	"	1.91	0.20	2.11
5200	1-1/2" furring channels				
5210	12" o.c.	SF	2.33	0.62	2.95
5220	16" o.c.	"	2.17	0.47	2.64
5230	24" o.c.	"	2.03	0.31	2.34

		UNIT	LABOR	MAT.	TOTAL
09205.10	**GYPSUM LATH**				
1070	Gypsum lath, 1/2" thick				
1090	Clipped	SY	3.62	5.76	9.38
1110	Nailed	"	4.07	5.76	9.83
09205.20	**METAL LATH**				
0960	Diamond expanded, galvanized				
0980	2.5 lb., on walls				
1010	Nailed	SY	8.15	3.83	11.98
1030	Wired	"	9.32	3.83	13.15
1040	On ceilings				
1050	Nailed	SY	9.32	3.83	13.15
1070	Wired	"	10.75	3.83	14.58
2230	Stucco lath				
2240	1.8 lb.	SY	8.15	4.51	12.66
2300	3.6 lb.	"	8.15	5.06	13.21
2310	Paper backed				
2320	Minimum	SY	6.52	3.50	10.02
2400	Maximum	"	9.32	5.65	14.97
09205.60	**PLASTER ACCESSORIES**				
0120	Expansion joint, 3/4", 26 ga., galv.	LF	1.63	1.48	3.11
2000	Plaster corner beads, 3/4", galvanized	"	1.86	0.41	2.27
2020	Casing bead, expanded flange, galvanized	"	1.63	0.56	2.19
2100	Expanded wing, 1-1/4" wide, galvanized	"	1.63	0.66	2.29
2500	Joint clips for lath	EA	0.32	0.17	0.49
2580	Metal base, galvanized, 2-1/2" high	LF	2.17	0.75	2.92
2600	Stud clips for gypsum lath	EA	0.32	0.17	0.49
2700	Tie wire galvanized, 18 ga., 25 lb. hank	"			47.00
8000	Sound deadening board, 1/4"	SF	1.08	0.31	1.39
09210.10	**PLASTER**				
0980	Gypsum plaster, trowel finish, 2 coats				
1000	Ceilings	SY	19.00	4.30	23.30
1020	Walls	"	17.75	4.30	22.05
1030	3 coats				
1040	Ceilings	SY	26.50	5.96	32.46
1060	Walls	"	23.25	5.96	29.21
7000	On columns, add to installation, 50%	"			
7020	Chases, fascia, and soffits, add to installation, 50%	"			
7040	Beams, add to installation, 50%	"			
09220.10	**PORTLAND CEMENT PLASTER**				
2980	Stucco, portland, gray, 3 coat, 1" thick				
3000	Sand finish	SY	26.50	8.03	34.53
3020	Trowel finish	"	27.50	8.03	35.53
3030	White cement				
3040	Sand finish	SY	27.50	9.17	36.67
3060	Trowel finish	"	30.25	9.17	39.42
3980	Scratch coat				
4000	For ceramic tile	SY	6.07	2.91	8.98
4020	For quarry tile	"	6.07	2.91	8.98
5000	Portland cement plaster				
5020	2 coats, 1/2"	SY	12.25	5.78	18.03
5040	3 coats, 7/8"	"	15.25	6.90	22.15

		UNIT	LABOR	MAT.	TOTAL
09250.10	**GYPSUM BOARD**				
0220	1/2", clipped to				
0240	Metal furred ceiling	SF	0.72	0.42	1.14
0260	Columns and beams	"	1.63	0.38	2.01
0270	Walls	"	0.65	0.38	1.03
0280	Nailed or screwed to				
0290	Wood or metal framed ceiling	SF	0.65	0.38	1.03
0300	Columns and beams	"	1.44	0.38	1.82
0400	Walls	"	0.59	0.38	0.97
1000	5/8", clipped to				
1020	Metal furred ceiling	SF	0.81	0.42	1.23
1040	Columns and beams	"	1.81	0.42	2.23
1060	Walls	"	0.72	0.42	1.14
1070	Nailed or screwed to				
1080	Wood or metal framed ceiling	SF	0.81	0.42	1.23
1100	Columns and beams	"	1.81	0.42	2.23
1120	Walls	"	0.72	0.42	1.14
1122	Vinyl faced, clipped to metal studs				
1124	1/2"	SF	0.81	1.19	2.00
1126	5/8"	"	0.81	1.13	1.94
1130	Add for				
1140	Fire resistant	SF			0.12
1180	Water resistant	"			0.19
1200	Water and fire resistant	"			0.24
1220	Taping and finishing joints				
1222	Minimum	SF	0.43	0.04	0.47
1224	Average	"	0.54	0.07	0.61
1226	Maximum	"	0.65	0.10	0.75
5020	Casing bead				
5022	Minimum	LF	1.86	0.16	2.02
5024	Average	"	2.17	0.18	2.35
5026	Maximum	"	3.26	0.22	3.48
5040	Corner bead				
5042	Minimum	LF	1.86	0.18	2.04
5044	Average	"	2.17	0.22	2.39
5046	Maximum	"	3.26	0.27	3.53
09310.10	**CERAMIC TILE**				
0980	Glazed wall tile, 4-1/4" x 4-1/4"				
1000	Minimum	SF	4.44	2.32	6.76
1020	Average	"	5.18	3.68	8.86
1040	Maximum	"	6.22	13.25	19.47
1042	6" x 6"				
1044	Minimum	SF	3.88	1.65	5.53
1046	Average	"	4.44	2.22	6.66
1048	Maximum	"	5.18	2.77	7.95
2960	Base, 4-1/4" high				
2980	Minimum	LF	7.77	4.47	12.24
3000	Average	"	7.77	5.20	12.97
3040	Maximum	"	7.77	6.87	14.64
3042	Glazed moldings and trim, 12" x 12"				
3044	Minimum	LF	6.22	2.46	8.68
3046	Average	"	6.22	3.75	9.97

DIVISION # 09 FINISHES

		UNIT	LABOR	MAT.	TOTAL
09310.10	**CERAMIC TILE, Cont'd...**				
3048	Maximum	LF	6.22	5.04	11.26
6100	Unglazed floor tile				
6120	Portland cem., cushion edge, face mtd				
6140	1" x 1"	SF	5.65	8.87	14.52
6150	2" x 2"	"	5.18	9.38	14.56
6162	4" x 4"	"	5.18	8.73	13.91
6164	6" x 6"	"	4.44	3.12	7.56
6166	12" x 12"	"	3.88	2.75	6.63
6168	16" x 16"	"	3.45	2.38	5.83
6170	18" x 18"	"	3.11	2.31	5.42
6200	Adhesive bed, with white grout				
6220	1" x 1"	SF	5.65	7.38	13.03
6230	2" x 2"	"	5.18	7.81	12.99
6260	4" x 4"	"	5.18	7.81	12.99
6262	6" x 6"	"	4.44	2.60	7.04
6264	12" x 12"	"	3.88	2.28	6.16
6266	16" x 16"	"	3.45	1.98	5.43
6268	18" x 18"	"	3.11	1.92	5.03
6300	Organic adhesive bed, thin set, back mounted				
6320	1" x 1"	SF	5.65	7.38	13.03
6350	2" x 2"	"	5.18	8.59	13.77
6360	For group 2 colors, add to material, 10%				
6370	For group 3 colors, add to material, 20%				
6380	For abrasive surface, add to material, 25%				
8990	Ceramic accessories				
9000	Towel bar, 24" long				
9004	Average	EA	31.00	22.25	53.25
9020	Soap dish				
9024	Average	EA	52.00	11.50	63.50
09330.10	**QUARRY TILE**				
1060	Floor				
1080	4 x 4 x 1/2"	SF	8.29	6.57	14.86
1120	6 x 6 x 3/4"	"	7.77	7.99	15.76
1200	Wall, applied to 3/4" portland cement bed				
1220	4 x 4 x 1/2"	SF	12.50	5.84	18.34
1240	6 x 6 x 3/4"	"	10.25	6.53	16.78
1320	Cove base				
1330	5 x 6 x 1/2" straight top	LF	10.25	6.66	16.91
1340	6 x 6 x 3/4" round top	"	10.25	6.18	16.43
1360	Stair treads 6 x 6 x 3/4"	"	15.50	9.13	24.63
1380	Window sill 6 x 8 x 3/4"	"	12.50	8.33	20.83
1400	For abrasive surface, add to material, 25%				
09410.10	**TERRAZZO**				
1100	Floors on concrete, 1-3/4" thick, 5/8" topping				
1120	Gray cement	SF	8.68	5.67	14.35
1140	White cement	"	8.68	5.92	14.60
1200	Sand cushion, 3" thick, 5/8" top, 1/4"				
1220	Gray cement	SF	10.00	6.69	16.69
1240	White cement	"	10.00	6.96	16.96
1260	Monolithic terrazzo, 3-1/2" base slab, 5/8" topping	"	7.59	4.75	12.34
1280	Terrazzo wainscot, cast-in-place, 1/2" thick	"	15.25	5.77	21.02

		UNIT	LABOR	MAT.	TOTAL
09410.10	**TERRAZZO, Cont'd...**				
1300	Base, cast-in-place, terrazzo cove type, 6" high	LF	8.68	7.26	15.94
1320	Curb, cast-in-place, 6" wide x 6" high, polished top	"	30.25	6.60	36.85
1400	For venetian type terrazzo, add to material, 10%				
1420	For abrasive heavy duty terrazzo, add to material, 15%				
1480	Divider strips				
1500	Zinc	LF			1.43
1510	Brass	"			2.66
09510.10	**CEILINGS AND WALLS**				
1520	Acoustical panels, suspension system not included				
1540	Fiberglass panels				
1550	5/8" thick				
1560	2' x 2'	SF	0.93	1.61	2.54
1580	2' x 4'	"	0.72	1.34	2.06
1590	3/4" thick				
1600	2' x 2'	SF	0.93	2.14	3.07
1620	2' x 4'	"	0.72	2.07	2.79
1640	Glass cloth faced fiberglass panels				
1660	3/4" thick	SF	1.08	3.05	4.13
1680	1" thick	"	1.08	3.41	4.49
1700	Mineral fiber panels				
1710	5/8" thick				
1720	2' x 2'	SF	0.93	1.37	2.30
1740	2' x 4'	"	0.72	1.37	2.09
1750	3/4" thick				
1760	2' x 2'	SF	0.93	2.14	3.07
1780	2' x 4'	"	0.72	2.07	2.79
3000	Acoustical tiles, suspension system not included				
3020	Fiberglass tile, 12" x 12"				
3040	5/8" thick	SF	1.18	2.00	3.18
3060	3/4" thick	"	1.44	2.32	3.76
3080	Glass cloth faced fiberglass tile				
3100	3/4" thick	SF	1.44	3.73	5.17
3120	3" thick	"	1.63	4.17	5.80
3140	Mineral fiber tile, 12" x 12"				
3150	5/8" thick				
3160	Standard	SF	1.30	1.05	2.35
3180	Vinyl faced	"	1.30	2.08	3.38
3190	3/4" thick				
3195	Standard	SF	1.30	1.53	2.83
3200	Vinyl faced	"	1.30	2.66	3.96
5500	Ceiling suspension systems				
5505	T-bar system				
5510	2' x 4'	SF	0.65	1.15	1.80
5520	2' x 2'	"	0.72	1.25	1.97
5530	Concealed Z-bar suspension system, 12" module	"	1.08	1.18	2.26
5550	For 1-1/2" carrier channels, 4' o.c., add	"			0.38
5560	Carrier channel for recessed light fixtures	"			0.69
09550.10	**WOOD FLOORING**				
0100	Wood strip flooring, unfinished				
1000	Fir floor				
1010	C and better				

DIVISION # 09 FINISHES

		UNIT	LABOR	MAT.	TOTAL
09550.10	**WOOD FLOORING, Cont'd...**				
1020	Vertical grain	SF	2.17	3.52	5.69
1040	Flat grain	"	2.17	4.40	6.57
1060	Oak floor				
1080	Minimum	SF	3.10	3.72	6.82
1100	Average	"	3.10	5.13	8.23
1120	Maximum	"	3.10	7.42	10.52
1340	Added costs				
1350	For factory finish, add to material, 10%				
1355	For random width floor, add to total, 20%				
1360	For simulated pegs, add to total, 10%				
3000	Gym floor, 2 ply felt, 25/32" maple, finished, in mastic	SF	3.62	8.54	12.16
3020	Over wood sleepers	"	4.07	8.69	12.76
9020	Finishing, sand, fill, finish, and wax	"	1.63	0.66	2.29
9100	Refinish sand, seal, and 2 coats of polyurethane	"	2.17	1.16	3.33
9540	Clean and wax floors	"	0.32	0.24	0.56
09550.20	**BAMBOO FLOORING**				
0010	Vertical, Carbonized Medium, 3' vertical grain	SF	2.17	5.85	8.02
0020	Natural	"	2.17	5.75	7.92
0030	3' horizontal grain	"	2.17	5.85	8.02
0040	Natural	"	2.17	5.75	7.92
0050	3' Stained	"	2.17	5.64	7.81
0060	6' spice	"	2.17	5.64	7.81
0070	3' stained, butterscotch	"	2.17	5.64	7.81
0080	6' tiger	"	2.17	5.64	7.81
0090	3' stained, Irish moss	"	2.17	5.64	7.81
0100	Vice-lock, 12 mm., laminate flooring, maple	"	2.17	3.72	5.89
0101	Oak	"	2.17	3.20	5.37
0102	Pine	"	2.17	4.26	6.43
0103	Espresso	"	2.17	4.26	6.43
0104	Standard, hard maple	"	2.17	3.46	5.63
0105	Cherry	"	2.17	3.20	5.37
0106	Oak	"	2.17	2.13	4.30
0107	Walnut	"	2.17	2.13	4.30
0108	South pacific vice-lock, 12 mm, brazilian cherry	"	2.17	4.79	6.96
0109	Maple	"	2.17	4.26	6.43
0110	Teak	"	2.17	4.53	6.70
09630.10	**UNIT MASONRY FLOORING**				
1000	Clay brick				
1020	9 x 4-1/2 x 3" thick				
1040	Glazed	SF	5.43	8.03	13.46
1060	Unglazed	"	5.43	7.70	13.13
1070	8 x 4 x 3/4" thick				
1080	Glazed	SF	5.67	7.26	12.93
1100	Unglazed	"	5.67	6.93	12.60
1140	For herringbone pattern, add to labor, 15%				
09660.10	**RESILIENT TILE FLOORING**				
1020	Solid vinyl tile, 1/8" thick, 12" x 12"				
1040	Marble patterns	SF	1.63	4.66	6.29
1060	Solid colors	"	1.63	6.05	7.68
1080	Travertine patterns	"	1.63	6.79	8.42
2000	Conductive resilient flooring, vinyl tile				

DIVISION # 09 FINISHES

		UNIT	LABOR	MAT.	TOTAL
09660.10	**RESILIENT TILE FLOORING, Cont'd...**				
2040	1/8" thick, 12" x 12"	SF	1.86	7.38	9.24
09665.10	**RESILIENT SHEET FLOORING**				
0980	Vinyl sheet flooring				
1000	Minimum	SF	0.65	3.83	4.48
1002	Average	"	0.77	6.19	6.96
1004	Maximum	"	1.08	10.50	11.58
1020	Cove, to 6"	LF	1.30	2.28	3.58
2000	Fluid applied resilient flooring				
2020	Polyurethane, poured in place, 3/8" thick	SF	5.43	10.50	15.93
6200	Vinyl sheet goods, backed				
6220	0.070" thick	SF	0.81	3.90	4.71
6260	0.125" thick	"	0.81	6.98	7.79
6280	0.250" thick	"	0.81	8.03	8.84
09678.10	**RESILIENT BASE AND ACCESSORIES**				
1000	Wall base, vinyl				
1130	4" high	LF	2.17	1.28	3.45
1140	6" high	"	2.17	1.74	3.91
09682.10	**CARPET PADDING**				
1000	Carpet padding				
1005	Foam rubber, waffle type, 0.3" thick	SY	3.26	6.74	10.00
1010	Jute padding				
1022	Average	SY	3.26	5.95	9.21
1030	Sponge rubber cushion				
1042	Average	SY	3.26	7.22	10.48
1050	Urethane cushion, 3/8" thick				
1062	Average	SY	3.26	6.32	9.58
09685.10	**CARPET**				
0990	Carpet, acrylic				
1000	24 oz., light traffic	SY	7.24	17.75	24.99
1020	28 oz., medium traffic	"	7.24	21.25	28.49
2010	Nylon				
2020	15 oz., light traffic	SY	7.24	24.50	31.74
2040	28 oz., medium traffic	"	7.24	32.00	39.24
2110	Nylon				
2120	28 oz., medium traffic	SY	7.24	30.50	37.74
2140	35 oz., heavy traffic	"	7.24	37.25	44.49
2145	Wool				
2150	30 oz., medium traffic	SY	7.24	51.00	58.24
2160	36 oz., medium traffic	"	7.24	53.00	60.24
2180	42 oz., heavy traffic	"	7.24	71.00	78.24
3000	Carpet tile				
3020	Foam backed				
3022	Minimum	SF	1.30	4.07	5.37
3024	Average	"	1.44	4.71	6.15
3026	Maximum	"	1.63	7.47	9.10
8980	Clean and vacuum carpet				
9000	Minimum	SY	0.25	0.36	0.61
9020	Average	"	0.43	0.56	0.99
9040	Maximum	"	0.65	0.77	1.42

		UNIT	LABOR	MAT.	TOTAL
09905.10	**PAINTING PREPARATION**				
1000	Dropcloths				
1050	Minimum	SF	0.03	0.16	0.19
1100	Average	"	0.04	0.18	0.22
1150	Maximum	"	0.06	0.37	0.43
1200	Masking				
1250	Paper and tape				
1300	Minimum	LF	0.54	0.04	0.58
1350	Average	"	0.68	0.06	0.74
1400	Maximum	"	0.90	0.07	0.97
1450	Doors				
1500	Minimum	EA	6.80	0.05	6.85
1550	Average	"	9.07	0.06	9.13
1600	Maximum	"	12.00	0.07	12.07
1650	Windows				
1700	Minimum	EA	6.80	0.05	6.85
1750	Average	"	9.07	0.06	9.13
1800	Maximum	"	12.00	0.07	12.07
2000	Sanding				
2050	Walls and flat surfaces				
2100	Minimum	SF	0.36		0.36
2150	Average	"	0.45		0.45
2200	Maximum	"	0.54		0.54
2250	Doors and windows				
2300	Minimum	EA	9.07		9.07
2350	Average	"	13.50		13.50
2400	Maximum	"	18.25		18.25
2450	Trim				
2500	Minimum	LF	0.68		0.68
2550	Average	"	0.90		0.90
2600	Maximum	"	1.20		1.20
2650	Puttying				
2700	Minimum	SF	0.83	0.01	0.84
2750	Average	"	1.08	0.02	1.10
2800	Maximum	"	1.36	0.03	1.39
09910.05	**EXT. PAINTING, SITEWORK**				
3000	Concrete Block				
3020	Roller				
3040	First Coat				
3060	Minimum	SF	0.27	0.19	0.46
3080	Average	"	0.36	0.21	0.57
3100	Maximum	"	0.54	0.22	0.76
3120	Second Coat				
3140	Minimum	SF	0.22	0.19	0.41
3160	Average	"	0.30	0.21	0.51
3180	Maximum	"	0.45	0.22	0.67
3200	Spray				
3220	First Coat				
3240	Minimum	SF	0.15	0.15	0.30
3260	Average	"	0.18	0.17	0.35
3280	Maximum	"	0.20	0.18	0.38
3300	Second Coat				

		UNIT	LABOR	MAT.	TOTAL
09910.05	**EXT. PAINTING, SITEWORK, Cont'd...**				
3320	Minimum	SF	0.09	0.15	0.24
3340	Average	"	0.12	0.17	0.29
3360	Maximum	"	0.17	0.18	0.35
3500	Fences, Chain Link				
3700	Roller				
3720	First Coat				
3740	Minimum	SF	0.38	0.13	0.51
3760	Average	"	0.45	0.14	0.59
3780	Maximum	"	0.51	0.15	0.66
3800	Second Coat				
3820	Minimum	SF	0.22	0.13	0.35
3840	Average	"	0.27	0.14	0.41
3860	Maximum	"	0.34	0.15	0.49
3880	Spray				
3900	First Coat				
3920	Minimum	SF	0.17	0.10	0.27
3940	Average	"	0.19	0.11	0.30
3960	Maximum	"	0.22	0.13	0.35
3980	Second Coat				
4000	Minimum	SF	0.12	0.10	0.22
4060	Average	"	0.15	0.11	0.26
4080	Maximum	"	0.17	0.13	0.30
4200	Fences, Wood or Masonry				
4220	Brush				
4240	First Coat				
4260	Minimum	SF	0.57	0.19	0.76
4280	Average	"	0.68	0.21	0.89
4300	Maximum	"	0.90	0.22	1.12
4320	Second Coat				
4340	Minimum	SF	0.34	0.19	0.53
4360	Average	"	0.41	0.21	0.62
4380	Maximum	"	0.54	0.22	0.76
4400	Roller				
4420	First Coat				
4440	Minimum	SF	0.30	0.19	0.49
4460	Average	"	0.36	0.21	0.57
4480	Maximum	"	0.41	0.22	0.63
4500	Second Coat				
4520	Minimum	SF	0.20	0.19	0.39
4540	Average	"	0.25	0.21	0.46
4560	Maximum	"	0.34	0.22	0.56
4580	Spray				
4600	First Coat				
4620	Minimum	SF	0.19	0.15	0.34
4640	Average	"	0.24	0.17	0.41
4660	Maximum	"	0.34	0.18	0.52
4680	Second Coat				
4700	Minimum	SF	0.13	0.15	0.28
4760	Average	"	0.17	0.17	0.34
4780	Maximum	"	0.22	0.18	0.40

DIVISION # 09 FINISHES

	UNIT	LABOR	MAT.	TOTAL
09910.15 EXT. PAINTING, BUILDINGS				
1200 Decks, Wood, Stained				
1580 Spray				
1600 First Coat				
1620 Minimum	SF	0.17	0.13	0.30
1640 Average	"	0.18	0.14	0.32
1660 Maximum	"	0.20	0.15	0.35
1680 Second Coat				
1700 Minimum	SF	0.15	0.13	0.28
1720 Average	"	0.16	0.14	0.30
1740 Maximum	"	0.18	0.15	0.33
2520 Doors, Wood				
2540 Brush				
2560 First Coat				
2580 Minimum	SF	0.83	0.15	0.98
2600 Average	"	1.08	0.17	1.25
2620 Maximum	"	1.36	0.18	1.54
2640 Second Coat				
2660 Minimum	SF	0.68	0.15	0.83
2680 Average	"	0.77	0.17	0.94
2700 Maximum	"	0.90	0.18	1.08
3680 Siding, Wood				
3880 Spray				
3900 First Coat				
3920 Minimum	SF	0.18	0.13	0.31
3940 Average	"	0.19	0.14	0.33
3960 Maximum	"	0.20	0.15	0.35
3980 Second Coat				
4000 Minimum	SF	0.13	0.13	0.26
4020 Average	"	0.18	0.14	0.32
4040 Maximum	"	0.27	0.15	0.42
4440 Trim				
4460 Brush				
4480 First Coat				
4500 Minimum	LF	0.22	0.19	0.41
4520 Average	"	0.27	0.21	0.48
4540 Maximum	"	0.34	0.22	0.56
4560 Second Coat				
4580 Minimum	LF	0.17	0.19	0.36
4600 Average	"	0.22	0.21	0.43
4620 Maximum	"	0.34	0.22	0.56
4640 Walls				
4840 Spray				
4860 First Coat				
4880 Minimum	SF	0.08	0.11	0.19
4900 Average	"	0.10	0.13	0.23
4920 Maximum	"	0.13	0.14	0.27
4940 Second Coat				
4960 Minimum	SF	0.07	0.11	0.18
4980 Average	"	0.09	0.13	0.22
5000 Maximum	"	0.12	0.14	0.26
5020 Windows				
5040 Brush				

		UNIT	LABOR	MAT.	TOTAL
09910.15	**EXT. PAINTING, BUILDINGS, Cont'd...**				
5060	First Coat				
5080	Minimum	SF	0.90	0.13	1.03
5100	Average	"	1.08	0.14	1.22
5120	Maximum	"	1.36	0.15	1.51
5140	Second Coat				
5160	Minimum	SF	0.77	0.13	0.90
5180	Average	"	0.90	0.14	1.04
5200	Maximum	"	1.08	0.15	1.23
09910.35	**INT. PAINTING, BUILDINGS**				
1380	Cabinets and Casework				
1400	Brush				
1420	First Coat				
1440	Minimum	SF	0.54	0.19	0.73
1460	Average	"	0.60	0.21	0.81
1480	Maximum	"	0.68	0.22	0.90
1500	Second Coat				
1520	Minimum	SF	0.45	0.19	0.64
1540	Average	"	0.49	0.21	0.70
1560	Maximum	"	0.54	0.22	0.76
1580	Spray				
1600	First Coat				
1620	Minimum	SF	0.27	0.15	0.42
1640	Average	"	0.32	0.17	0.49
1660	Maximum	"	0.38	0.18	0.56
1680	Second Coat				
1700	Minimum	SF	0.21	0.15	0.36
1720	Average	"	0.23	0.17	0.40
1740	Maximum	"	0.30	0.18	0.48
2520	Doors, Wood				
2540	Brush				
2560	First Coat				
2580	Minimum	SF	0.77	0.19	0.96
2600	Average	"	0.98	0.21	1.19
2620	Maximum	"	1.20	0.22	1.42
2640	Second Coat				
2660	Minimum	SF	0.60	0.14	0.74
2680	Average	"	0.68	0.15	0.83
2700	Maximum	"	0.77	0.17	0.94
2720	Spray				
2740	First Coat				
2760	Minimum	SF	0.16	0.14	0.30
2780	Average	"	0.19	0.15	0.34
2800	Maximum	"	0.24	0.17	0.41
2820	Second Coat				
2840	Minimum	SF	0.12	0.14	0.26
2860	Average	"	0.14	0.15	0.29
2880	Maximum	"	0.17	0.17	0.34
3900	Trim				
3920	Brush				
3940	First Coat				
3960	Minimum	LF	0.21	0.19	0.40

DIVISION # 09 FINISHES

		UNIT	LABOR	MAT.	TOTAL
09910.35	**INT. PAINTING, BUILDINGS, Cont'd...**				
3980	Average	LF	0.24	0.21	0.45
4000	Maximum	"	0.30	0.22	0.52
4020	Second Coat				
4040	Minimum	LF	0.16	0.19	0.35
4060	Average	"	0.20	0.21	0.41
4080	Maximum	"	0.30	0.22	0.52
4100	Walls				
4120	Roller				
4140	First Coat				
4160	Minimum	SF	0.19	0.15	0.34
4180	Average	"	0.20	0.17	0.37
4200	Maximum	"	0.22	0.18	0.40
4220	Second Coat				
4240	Minimum	SF	0.17	0.15	0.32
4260	Average	"	0.18	0.17	0.35
4280	Maximum	"	0.20	0.18	0.38
4300	Spray				
4320	First Coat				
4340	Minimum	SF	0.08	0.13	0.21
4360	Average	"	0.10	0.14	0.24
4380	Maximum	"	0.13	0.15	0.28
4400	Second Coat				
4420	Minimum	SF	0.07	0.13	0.20
4440	Average	"	0.09	0.14	0.23
4460	Maximum	"	0.12	0.15	0.27
09955.10	**WALL COVERING**				
0900	Vinyl wall covering				
1000	Medium duty	SF	0.77	1.10	1.87
1010	Heavy duty	"	0.90	2.26	3.16

Design Cost Data™ **DCD**

TABLE OF CONTENTS   PAGE

		UNIT	LABOR	MAT.	TOTAL
10110.10	**CHALKBOARDS**				
1020	Chalkboard, metal frame, 1/4" thick				
1040	48"x60"	EA	65.00	480	545
1060	48"x96"	"	72.00	660	732
1080	48"x144"	"	82.00	880	962
1100	48"x192"	"	93.00	1,190	1,283
1110	Liquid chalkboard				
1120	48"x60"	EA	65.00	640	705
1140	48"x96"	"	72.00	820	892
1160	48"x144"	"	82.00	1,220	1,302
1180	48"x192"	"	93.00	1,400	1,493
1200	Map rail, deluxe	LF	3.26	7.97	11.23
10165.10	**TOILET PARTITIONS**				
0100	Toilet partition, plastic laminate				
0120	Ceiling mounted	EA	220	1,180	1,400
0140	Floor mounted	"	160	780	940
0150	Metal				
0165	Ceiling mounted	EA	220	810	1,030
0180	Floor mounted	"	160	770	930
0190	Wheelchair partition, plastic laminate				
0200	Ceiling mounted	EA	220	1,770	1,990
0210	Floor mounted	"	160	1,550	1,710
0215	Painted metal				
0220	Ceiling mounted	EA	220	1,260	1,480
0230	Floor mounted	"	160	1,150	1,310
1980	Urinal screen, plastic laminate				
2000	Wall hung	EA	82.00	550	632
2100	Floor mounted	"	82.00	490	572
2120	Porcelain enameled steel, floor mounted	"	82.00	630	712
2140	Painted metal, floor mounted	"	82.00	420	502
2160	Stainless steel, floor mounted	"	82.00	800	882
5000	Metal toilet partitions				
5020	Front door and side divider, floor mounted				
5040	Porcelain enameled steel	EA	160	1,300	1,460
5060	Painted steel	"	160	760	920
5080	Stainless steel	"	160	1,890	2,050
10185.10	**SHOWER STALLS**				
1000	Shower receptors				
1010	Precast, terrazzo				
1020	32" x 32"	EA	60.00	690	750
1040	32" x 48"	"	72.00	720	792
1050	Concrete				
1060	32" x 32"	EA	60.00	280	340
1080	48" x 48"	"	79.00	310	389
1100	Shower door, trim and hardware				
1130	Porcelain enameled steel, flush	EA	72.00	560	632
1140	Baked enameled steel, flush	"	72.00	330	402
1150	Aluminum, tempered glass, 48" wide, sliding	"	89.00	690	779
1161	Folding	"	89.00	670	759
5400	Shower compartment, precast concrete receptor				
5420	Single entry type				
5440	Porcelain enameled steel	EA	720	2,380	3,100

		UNIT	LABOR	MAT.	TOTAL
10185.10	**SHOWER STALLS, Cont'd...**				
5460	Baked enameled steel	EA	720	2,280	3,000
5480	Stainless steel	"	720	2,190	2,910
5500	Double entry type				
5520	Porcelain enameled steel	EA	890	4,250	5,140
5540	Baked enameled steel	"	890	2,900	3,790
5560	Stainless steel	"	890	4,690	5,580
10210.10	**VENTS AND WALL LOUVERS**				
1230	Grilles and louvers				
2040	Fixed type louvers				
2060	4 through 10 sf	SF	12.00	34.50	46.50
2080	Over 10 sf	"	9.00	40.75	49.75
2090	Movable type louvers				
2220	4 through 10 sf	SF	12.00	40.75	52.75
2240	Over 10 sf	"	9.00	45.00	54.00
2260	Aluminum louvers				
4000	Residential use, fixed type, with screen				
4020	8" x 8"	EA	36.00	20.75	56.75
4060	12" x 12"	"	36.00	23.00	59.00
4080	12" x 18"	"	36.00	27.50	63.50
4100	14" x 24"	"	36.00	39.50	75.50
4120	18" x 24"	"	36.00	44.50	80.50
4140	30" x 24"	"	40.00	61.00	101
10290.10	**PEST CONTROL**				
1000	Termite control				
1010	Under slab spraying				
1020	Minimum	SF	0.12	1.21	1.33
1040	Average	"	0.25	1.21	1.46
1120	Maximum	"	0.51	1.73	2.24
10350.10	**FLAGPOLES**				
2020	Installed in concrete base				
2030	Fiberglass				
2040	25' high	EA	430	1,640	2,070
2080	50' high	"	1,090	4,330	5,420
2100	Aluminum				
2120	25' high	EA	430	1,590	2,020
2140	50' high	"	1,090	3,150	4,240
2160	Bonderized steel				
2180	25' high	EA	500	1,780	2,280
2200	50' high	"	1,300	3,560	4,860
2220	Freestanding tapered, fiberglass				
2240	30' high	EA	470	1,950	2,420
2260	40' high	"	590	2,540	3,130
2280	50' high	"	650	6,460	7,110
2300	60' high	"	770	6,910	7,680
2400	Wall mounted, with collar, brushed aluminum finish				
2420	15' long	EA	330	1,510	1,840
2440	18' long	"	330	1,710	2,040
2460	20' long	"	340	1,860	2,200
2480	24' long	"	380	2,000	2,380
2500	Outrigger, wall, including base				
2520	10' long	EA	430	1,530	1,960

		UNIT	LABOR	MAT.	TOTAL
10350.10	**FLAGPOLES, Cont'd...**				
2540	20' long	EA	540	2,030	2,570
10400.10	**IDENTIFYING DEVICES**				
1000	Directory and bulletin boards				
1020	Open face boards				
1040	Chrome plated steel frame	SF	32.50	37.75	70.25
1060	Aluminum framed	"	32.50	65.00	97.50
1080	Bronze framed	"	32.50	84.00	117
1100	Stainless steel framed	"	32.50	120	153
1140	Tack board, aluminum framed	"	32.50	26.50	59.00
1160	Visual aid board, aluminum framed	"	32.50	26.50	59.00
1200	Glass encased boards, hinged and keyed				
1210	Aluminum framed	SF	82.00	140	222
1220	Bronze framed	"	82.00	160	242
1230	Stainless steel framed	"	82.00	210	292
1240	Chrome plated steel framed	"	82.00	230	312
2020	Metal plaque				
2040	Cast bronze	SF	54.00	650	704
2060	Aluminum	"	54.00	370	424
2080	Metal engraved plaque				
2100	Porcelain steel	SF	54.00	780	834
2120	Stainless steel	"	54.00	620	674
2140	Brass	"	54.00	920	974
2160	Aluminum	"	54.00	570	624
2200	Metal built-up plaque				
2220	Bronze	SF	65.00	700	765
2240	Copper and bronze	"	65.00	620	685
2260	Copper and aluminum	"	65.00	680	745
2280	Metal nameplate plaques				
2300	Cast bronze	SF	40.75	690	731
2320	Cast aluminum	"	40.75	510	551
2330	Engraved, 1-1/2" x 6"				
2340	Bronze	EA	40.75	290	331
2360	Aluminum	"	40.75	220	261
2440	Letters, on masonry, aluminum, satin finish				
2450	1/2" thick				
2460	2" high	EA	26.00	27.00	53.00
2480	4" high	"	32.50	40.50	73.00
2500	6" high	"	36.25	54.00	90.25
2510	3/4" thick				
2520	8" high	EA	40.75	81.00	122
2540	10" high	"	46.50	94.00	141
2550	1" thick				
2560	12" high	EA	54.00	110	164
2580	14" high	"	65.00	120	185
2600	16" high	"	82.00	150	232
2620	For polished aluminum add, 15%				
2640	For clear anodized aluminum add, 15%				
2660	For colored anodic aluminum add, 30%				
2680	For profiled and color enameled letters add, 50%				
2700	Cast bronze, satin finish letters				
2710	3/8" thick				

		UNIT	LABOR	MAT.	TOTAL
10400.10	**IDENTIFYING DEVICES, Cont'd...**				
2720	2" high	EA	26.00	32.75	58.75
2740	4" high	"	32.50	49.25	81.75
2760	1/2" thick, 6" high	"	36.25	67.00	103
2780	5/8" thick, 8" high	"	40.75	100	141
2785	1" thick				
2790	10" high	EA	46.50	120	167
2800	12" high	"	54.00	150	204
2820	14" high	"	65.00	190	255
2840	16" high	"	82.00	270	352
3000	Interior door signs, adhesive, flexible				
3060	2" x 8"	EA	12.75	25.25	38.00
3080	4" x 4"	"	12.75	26.75	39.50
3100	6" x 7"	"	12.75	33.75	46.50
3120	6" x 9"	"	12.75	43.00	55.75
3140	10" x 9"	"	12.75	56.00	68.75
3160	10" x 12"	"	12.75	73.00	85.75
3200	Hard plastic type, no frame				
3220	3" x 8"	EA	12.75	56.00	68.75
3240	4" x 4"	"	12.75	56.00	68.75
3260	4" x 12"	"	12.75	60.00	72.75
3280	Hard plastic type, with frame				
3300	3" x 8"	EA	12.75	170	183
3320	4" x 4"	"	12.75	130	143
3340	4" x 12"	"	12.75	210	223
10450.10	**CONTROL**				
1020	Access control, 7' high, indoor or outdoor impenetrability				
1040	Remote or card control, type B	EA	890	1,850	2,740
1060	Free passage, type B	"	890	1,500	2,390
1080	Remote or card control, type AA	"	890	2,950	3,840
1100	Free passage, type AA	"	890	2,670	3,560
10500.10	**LOCKERS**				
0080	Locker bench, floor mounted, laminated maple				
0100	4'	EA	54.00	390	444
0120	6'	"	54.00	550	604
0130	Wardrobe locker, 12" x 60" x 15", baked on enamel				
0140	1-tier	EA	32.50	420	453
0160	2-tier	"	32.50	450	483
0180	3-tier	"	34.25	500	534
0200	4-tier	"	34.25	540	574
0240	12" x 72" x 15", baked on enamel				
0260	1-tier	EA	32.50	360	393
0280	2-tier	"	32.50	430	463
0300	4-tier	"	34.25	530	564
0320	5-tier	"	34.25	530	564
1200	15" x 60" x 15", baked on enamel				
1220	1-tier	EA	32.50	470	503
1240	4-tier	"	34.25	510	544
10520.10	**FIRE PROTECTION**				
1000	Portable fire extinguishers				
1020	Water pump tank type				
1030	2.5 gal.				

		UNIT	LABOR	MAT.	TOTAL
10520.10	**FIRE PROTECTION, Cont'd...**				
1040	Red enameled galvanized	EA	34.00	150	184
1060	Red enameled copper	"	34.00	220	254
1080	Polished copper	"	34.00	290	324
1200	Carbon dioxide type, red enamel steel				
1210	Squeeze grip with hose and horn				
1220	2.5 lb	EA	34.00	220	254
1240	5 lb	"	39.25	320	359
1260	10 lb	"	51.00	330	381
1280	15 lb	"	64.00	370	434
1300	20 lb	"	64.00	460	524
1310	Wheeled type				
1320	125 lb	EA	100	4,050	4,150
1340	250 lb	"	100	5,120	5,220
1360	500 lb	"	100	6,610	6,710
1400	Dry chemical, pressurized type				
1405	Red enameled steel				
1410	2.5 lb	EA	34.00	70.00	104
1430	5 lb	"	39.25	96.00	135
1440	10 lb	"	51.00	200	251
1450	20 lb	"	64.00	260	324
1460	30 lb	"	64.00	320	384
1480	Chrome plated steel, 2.5 lb	"	34.00	300	334
10550.10	**POSTAL SPECIALTIES**				
1500	Mail chutes				
1520	Single mail chute				
1530	Finished aluminum	LF	160	860	1,020
1540	Bronze	"	160	1,200	1,360
1560	Single mail chute receiving box				
1580	Finished aluminum	EA	330	1,280	1,610
1600	Bronze	"	330	1,540	1,870
1620	Twin mail chute, double parallel				
1630	Finished aluminum	FLR	330	1,890	2,220
1640	Bronze	"	330	2,470	2,800
10670.10	**SHELVING**				
0980	Shelving, enamel, closed side and back, 12" x 36"				
1000	5 shelves	EA	110	320	430
1020	8 shelves	"	140	350	490
1030	Open				
1040	5 shelves	EA	110	160	270
1060	8 shelves	"	140	180	320
2000	Metal storage shelving, baked enamel				
2030	7 shelf unit, 72" or 84" high				
2050	12" shelf	LF	69.00	51.00	120
2080	24" shelf	"	82.00	110	192
2100	36" shelf	"	93.00	120	213
2200	4 shelf unit, 40" high				
2240	12" shelf	LF	59.00	68.00	127
2270	24" shelf	"	72.00	140	212
2300	3 shelf unit, 32" high				
2340	12" shelf	LF	34.25	54.00	88.25
2370	24" shelf	"	40.75	68.00	109

DIVISION # 10 SPECIALTIES

		UNIT	LABOR	MAT.	TOTAL
10670.10	**SHELVING, Cont'd...**				
2400	Single shelf unit, attached to masonry				
2420	12" shelf	LF	11.75	21.00	32.75
2450	24" shelf	"	14.25	29.75	44.00
2460	For stainless steel, add to material, 120%				
2470	For attachment to gypsum board, add to labor, 50%				
10800.10	**BATH ACCESSORIES**				
1050	Grab bar, 1-1/2" dia., stainless steel, wall mounted				
1060	24" long	EA	32.50	57.00	89.50
1080	36" long	"	34.25	64.00	98.25
1100	48" long	"	38.25	78.00	116
1130	1" dia., stainless steel				
1140	12" long	EA	28.25	35.50	63.75
1180	24" long	"	32.50	48.25	80.75
1220	36" long	"	36.25	64.00	100
1240	48" long	"	38.25	71.00	109
1300	Hand dryer, surface mounted, 110 volt	"	82.00	810	892
1320	Medicine cabinet, 16 x 22, baked enamel, lighted	"	26.00	160	186
1340	With mirror, lighted	"	43.50	230	274
1420	Mirror, 1/4" plate glass, up to 10 sf	SF	6.52	12.50	19.02
1430	Mirror, stainless steel frame				
1440	18"x24"	EA	21.75	95.00	117
1500	24"x30"	"	32.50	120	153
1520	24"x48"	"	54.00	170	224
1560	30"x30"	"	65.00	380	445
1600	48"x72"	"	110	740	850
1640	With shelf, 18"x24"	"	26.00	290	316
1820	Sanitary napkin dispenser, stainless steel	"	43.50	700	744
1830	Shower rod, 1" diameter				
1840	Chrome finish over brass	EA	32.50	250	283
1860	Stainless steel	"	32.50	180	213
1900	Soap dish, stainless steel, wall mounted	"	43.50	160	204
1910	Toilet tissue dispenser, stainless, wall mounted				
1920	Single roll	EA	16.25	79.00	95.25
1940	Double roll	"	18.75	150	169
1945	Towel dispenser, stainless steel				
1950	Flush mounted	EA	36.25	300	336
1960	Surface mounted	"	32.50	430	463
1970	Combination towel and waste receptacle	"	43.50	660	704
2000	Towel bar, stainless steel				
2020	18" long	EA	26.00	98.00	124
2040	24" long	"	29.75	130	160
2060	30" long	"	32.50	140	173
2070	36" long	"	36.25	150	186
2100	Waste receptacle, stainless steel, wall mounted	"	54.00	500	554

TABLE OF CONTENTS   PAGE

		UNIT	LABOR	MAT.	TOTAL
11010.10	**MAINTENANCE EQUIPMENT**				
1000	Vacuum cleaning system				
1010	3 valves				
1020	1.5 hp	EA	720	1,000	1,720
1030	2.5 hp	"	930	1,200	2,130
1040	5 valves	"	1,300	1,870	3,170
1060	7 valves	"	1,630	2,490	4,120
11060.10	**THEATER EQUIPMENT**				
1000	Roll out stage, steel frame, wood floor				
1020	Manual	SF	4.07	58.00	62.07
1040	Electric	"	6.52	55.00	61.52
1100	Portable stages				
1120	8" high	SF	3.26	22.50	25.76
1140	18" high	"	3.62	26.25	29.87
1160	36" high	"	3.83	30.75	34.58
1180	48" high	"	4.07	34.25	38.32
1300	Band risers				
1320	Minimum	SF	3.26	59.00	62.26
1340	Maximum	"	3.26	120	123
1400	Chairs for risers				
1420	Minimum	EA	2.32	770	772
1440	Maximum	"	2.32	1,230	1,232
2000	Theatre controlls				
2010	Fade console, 48 channel	EA	330	3,060	3,390
2020	Light control modules, 125 channels	"	660	8,700	9,360
2030	Dimmer module, stage-pin output	"	330	2,680	3,010
2040	6-Module pack, w/24 U-ground connectors	"	330	7,650	7,980
11090.10	**CHECKROOM EQUIPMENT**				
1000	Motorized checkroom equipment				
1020	No shelf system, 6'4" height				
1040	7'6" length	EA	650	5,420	6,070
1060	14'6" length	"	650	5,480	6,130
1080	28' length	"	650	6,600	7,250
1100	One shelf, 6'8" height				
1120	7'6" length	EA	650	6,640	7,290
1140	14'6" length	"	650	6,820	7,470
1160	28' length	"	650	8,200	8,850
11110.10	**LAUNDRY EQUIPMENT**				
1000	High capacity, heavy duty				
1020	Washer extractors				
1030	135 lb				
1040	Standard	EA	540	37,190	37,730
1060	Pass through	"	540	42,730	43,270
1070	200 lb				
1080	Standard	EA	540	45,670	46,210
1100	Pass through	"	540	55,450	55,990
1120	110 lb dryer	"	540	13,990	14,530
11161.10	**LOADING DOCK EQUIPMENT**				
0080	Dock leveler, 10 ton capacity				
0100	6' x 8'	EA	650	5,760	6,410
0120	7' x 8'	"	650	6,610	7,260

DIVISION # 11 EQUIPMENT

11170.10	WASTE HANDLING	UNIT	LABOR	MAT.	TOTAL
1500	Industrial compactor				
1520	1 c.y.	EA	740	14,250	14,990
1540	3 c.y.	"	950	22,230	23,180
1560	5 c.y.	"	1,330	42,280	43,610
11170.20	**AERATION EQUIPMENT**				
0010	Surface spray/Vertical pump				
0020	1 hp Pump, 500 gpm.	EA	360	5,300	5,660
0030	5 hp Pump, 2000 gpm.	"	1,430	7,950	9,380
0040	Polycarbon, treatment container, 1,000 gallon capacity	"	2,850	4,230	7,080
0050	1,500 gallon	"	2,850	4,850	7,700
0060	2,000 gallon	"	3,560	6,750	10,310
0070	3,000 gallon	"	3,800	10,740	14,540
11400.10	**FOOD SERVICE EQUIPMENT**				
1000	Unit kitchens				
1020	30" compact kitchen				
1040	Refrigerator, with range, sink	EA	330	1,620	1,950
1060	Sink only	"	220	2,070	2,290
1080	Range only	"	170	1,680	1,850
1100	Cabinet for upper wall section	"	95.00	420	515
1120	Stainless shield, for rear wall	"	26.50	170	197
1140	Side wall	"	26.50	120	147
1200	42" compact kitchen				
1220	Refrigerator with range, sink	EA	370	1,980	2,350
1240	Sink only	"	330	1,080	1,410
1260	Cabinet for upper wall section	"	110	840	950
1280	Stainless shield, for rear wall	"	27.75	670	698
1290	Side wall	"	27.75	190	218
1600	Bake oven				
1610	Single deck				
1620	Minimum	EA	83.00	3,830	3,913
1640	Maximum	"	170	7,330	7,500
1650	Double deck				
1660	Minimum	EA	110	6,830	6,940
1680	Maximum	"	170	21,340	21,510
1685	Triple deck				
1690	Minimum	EA	110	24,220	24,330
1695	Maximum	"	220	43,210	43,430
1700	Convection type oven, electric, 40" x 45" x 57"				
1720	Minimum	EA	83.00	3,780	3,863
1740	Maximum	"	170	6,650	6,820
1800	Broiler, without oven, 69" x 26" x 39"				
1820	Minimum	EA	83.00	6,020	6,103
1840	Maximum	"	110	9,470	9,580
1900	Coffee urns, 10 gallons				
1920	Minimum	EA	220	4,580	4,800
1940	Maximum	"	330	5,180	5,510
2000	Fryer, with submerger				
2010	Single				
2020	Minimum	EA	130	1,880	2,010
2040	Maximum	"	220	5,160	5,380
2050	Double				

		UNIT	LABOR	MAT.	TOTAL
11400.10	**FOOD SERVICE EQUIPMENT, Cont'd...**				
2060	Minimum	EA	170	3,110	3,280
2080	Maximum	"	220	16,700	16,920
2100	Griddle, counter				
2110	3' long				
2120	Minimum	EA	110	2,570	2,680
2140	Maximum	"	130	5,300	5,430
2150	5' long				
2160	Minimum	EA	170	5,580	5,750
2180	Maximum	"	220	12,850	13,070
2200	Kettles, steam, jacketed				
2210	20 gallons				
2220	Minimum	EA	170	12,270	12,440
2240	Maximum	"	330	13,410	13,740
2300	Range				
2310	Heavy duty, single oven, open top				
2320	Minimum	EA	83.00	7,910	7,993
2340	Maximum	"	220	16,650	16,870
2350	Fry top				
2360	Minimum	EA	83.00	8,080	8,163
2380	Maximum	"	220	11,320	11,540
2400	Steamers, electric				
2410	27 kw				
2420	Minimum	EA	170	14,360	14,530
2440	Maximum	"	220	26,240	26,460
2450	18 kw				
2460	Minimum	EA	170	7,900	8,070
2480	Maximum	"	220	18,540	18,760
2500	Dishwasher, rack type				
2520	Single tank, 190 racks/hr	EA	330	21,050	21,380
2530	Double tank				
2540	234 racks/hr	EA	370	43,460	43,830
2560	265 racks/hr	"	440	51,950	52,390
2580	Dishwasher, automatic 100 meals/hr	"	220	16,900	17,120
2600	Disposals				
2620	100 gal/hr	EA	220	1,410	1,630
2640	120 gal/hr	"	230	1,640	1,870
2660	250 gal/hr	"	240	1,930	2,170
2680	Exhaust hood for dishwasher, gutter 4 sides				
2681	4'x4'x2'	EA	250	3,420	3,670
2690	4'x7'x2'	"	270	4,630	4,900
2900	Ice cube maker				
2910	50 lb per day				
2920	Minimum	EA	660	2,640	3,300
2940	Maximum	"	660	3,900	4,560
2950	500 lb per day				
2960	Minimum	EA	1,110	6,270	7,380
2970	Maximum	"	1,110	7,410	8,520
3100	Refrigerated cases				
3120	Dairy products				
3140	Multi-deck type	LF	44.25	1,490	1,534
3160	For rear sliding doors, add	"			280
3180	Delicatessen case, service deli				

		UNIT	LABOR	MAT.	TOTAL
11400.10	**FOOD SERVICE EQUIPMENT, Cont'd...**				
3190	Single deck	LF	330	1,060	1,390
3200	Multi-deck	"	420	1,210	1,630
3220	Meat case				
3230	Single deck	LF	390	910	1,300
3240	Multi-deck	"	420	1,070	1,490
3260	Produce case				
3270	Single deck	LF	390	1,060	1,450
3280	Multi-deck	"	420	1,130	1,550
3300	Bottle coolers				
3310	6' long				
3320	Minimum	EA	1,330	3,010	4,340
3340	Maximum	"	1,330	4,460	5,790
3345	10' long				
3350	Minimum	EA	2,210	3,900	6,110
3360	Maximum	"	2,210	7,670	9,880
3420	Frozen food cases				
3440	Chest type	LF	390	810	1,200
3460	Reach-in, glass door	"	420	1,120	1,540
3470	Island case, single	"	390	1,000	1,390
3480	Multi-deck	"	420	1,590	2,010
3500	Ice storage bins				
3520	500 lb capacity	EA	950	1,680	2,630
3530	1000 lb capacity	"	1,900	2,510	4,410
11450.10	**RESIDENTIAL EQUIPMENT**				
0300	Compactor, 4 to 1 compaction	EA	170	1,680	1,850
1310	Dishwasher, built-in				
1320	2 cycles	EA	330	830	1,160
1330	4 or more cycles	"	330	2,230	2,560
1340	Disposal				
1350	Garbage disposer	EA	220	230	450
1360	Heaters, electric, built-in				
1362	Ceiling type	EA	220	470	690
1364	Wall type				
1370	Minimum	EA	170	240	410
1374	Maximum	"	220	820	1,040
1400	Hood for range, 2-speed, vented				
1420	30" wide	EA	220	650	870
1440	42" wide	"	220	1,200	1,420
1460	Ice maker, automatic				
1480	30 lb per day	EA	95.00	2,200	2,295
1500	50 lb per day	"	330	2,790	3,120
2000	Ranges, electric				
2040	Built-in, 30", 1 oven	EA	220	2,410	2,630
2050	2 oven	"	220	2,790	3,010
2060	Countertop, 4 burner, standard	"	170	1,390	1,560
2070	With grill	"	170	3,480	3,650
2198	Freestanding, 21", 1 oven	"	220	1,250	1,470
2200	30", 1 oven	"	130	2,440	2,570
2220	2 oven	"	130	3,970	4,100
3600	Water softener				
3620	30 grains per gallon	EA	220	1,360	1,580

		UNIT	LABOR	MAT.	TOTAL
11450.10	**RESIDENTIAL EQUIPMENT, Cont'd...**				
3640	70 grains per gallon	EA	330	1,720	2,050
11600.10	**LABORATORY EQUIPMENT**				
1000	Cabinets, base				
1020	Minimum	LF	54.00	510	564
1040	Maximum	"	54.00	930	984
1080	Full storage, 7' high				
1100	Minimum	LF	54.00	490	544
1140	Maximum	"	54.00	930	984
1150	Wall				
1160	Minimum	LF	65.00	180	245
1200	Maximum	"	65.00	310	375
1220	Countertops				
1240	Minimum	SF	8.15	76.00	84.15
1260	Average	"	9.32	89.00	98.32
1280	Maximum	"	10.75	100	111
1300	Tables				
1320	Open underneath	SF	32.50	160	193
1330	Doors underneath	"	40.75	550	591
2000	Medical laboratory equipment				
2010	Analyzer				
2020	Chloride	EA	33.25	6,100	6,133
2060	Blood	"	55.00	33,540	33,595
2070	Bath, water, utility, countertop unit	"	66.00	1,260	1,326
2080	Hot plate, lab, countertop	"	60.00	480	540
2100	Stirrer	"	60.00	580	640
2120	Incubator, anaerobic, 23x23x36"	"	330	10,340	10,670
2140	Dry heat bath	"	110	1,120	1,230
2160	Incinerator, for sterilizing	"	6.64	770	777
2170	Meter, serum protein	"	8.30	1,190	1,198
2180	pH analog, general purpose	"	9.48	1,280	1,289
2190	Refrigerator, blood bank	"	110	9,780	9,890
2200	5.4 cf, undercounter type	"	110	6,640	6,750
2210	Refrigerator/freezer, 4.4 cf, undercounter type	"	110	1,260	1,370
2220	Sealer, impulse, free standing, 20x12x4"	"	22.25	700	722
2240	Timer, electric, 1-60 minutes, bench or wall mounted	"	37.00	270	307
2260	Glassware washer-dryer, undercounter	"	830	11,580	12,410
2300	Balance, torsion suspension, tabletop, 4.5 lb capacity	"	37.00	1,520	1,557
2340	Binocular microscope, with in-base illuminator	"	25.50	4,650	4,676
2400	Centrifuge, table model, 19x16x13"	"	26.50	1,780	1,807
2420	Clinical model, with four place head	"	14.75	1,930	1,945
11700.10	**MEDICAL EQUIPMENT**				
1000	Hospital equipment, lights				
1020	Examination, portable	EA	55.00	2,130	2,185
1200	Meters				
1220	Air flow meter	EA	37.00	110	147
1240	Oxygen flow meters	"	27.75	140	168
1900	Physical therapy				
1930	Chair, hydrotherapy	EA	10.75	820	831
1940	Diathermy, shortwave, portable, on casters	"	26.00	3,540	3,566
1950	Exercise bicycle, floor standing, 35" x 15"	"	21.75	3,440	3,462
1960	Hydrocollator, 4 pack, portable, 129 x 90 x 160"	"	9.32	620	629

		UNIT	LABOR	MAT.	TOTAL
11700.10	**MEDICAL EQUIPMENT, Cont'd...**				
1970	Lamp, infrared, mobile with variable heat control	EA	50.00	830	880
1980	Ultraviolet, base mounted	"	50.00	770	820
2070	Stimulator, galvanic-faradic, handheld	"	4.34	430	434
2080	Ultrasound stimulator, portable, 13x13x8"	"	5.43	3,680	3,685
2120	Whirlpool, 85 gallon	"	330	7,070	7,400
2141	65 gallon capacity	"	330	6,380	6,710
2200	Radiology				
2280	Radiographic table, motor driven tilting table	EA	6,520	62,350	68,870
2290	Fluoroscope image/tv system	"	13,050	103,440	116,490
2300	Processor for washing and drying radiographs				
2303	Water filter unit, 30" x 48-1/2" x 37-1/2"	EA	1,110	140	1,250
2400	Steam sterilizers				
2410	For heat and moisture stable materials	EA	66.00	6,020	6,086
2440	For fast drying after sterilization	"	83.00	7,790	7,873
2450	Compact unit	"	83.00	2,520	2,603
2460	Semi-automatic	"	330	2,980	3,310
2465	Floor loading				
2470	Single door	EA	550	88,280	88,830
2480	Double door	"	660	96,710	97,370
2490	Utensil washer, sanitizer	"	510	19,410	19,920
2500	Automatic washer/sterilizer	"	1,330	21,250	22,580
2510	16 x 16 x 26", including accessories	"	2,210	24,430	26,640
2520	Steam generator, elec., 10 kw to 180 kw	"	1,330	40,380	41,710
2540	Surgical scrub				
2560	Minimum	EA	220	2,130	2,350
2580	Maximum	"	220	12,330	12,550
2600	Gas sterilizers				
2620	Automatic, freestanding, 21x19x29"	EA	660	7,580	8,240
2720	Surgical lights, ceiling mounted				
2740	Minimum	EA	1,110	10,490	11,600
2760	Maximum	"	1,330	21,400	22,730
2800	Water stills				
2900	4 liters/hr	EA	220	4,750	4,970
2920	8 liters/hr	"	220	7,580	7,800
2940	19 liters/hr	"	550	15,590	16,140
3060	X-ray equipment				
3070	Mobile unit				
3080	Minimum	EA	330	13,880	14,210
3100	Maximum	"	660	26,880	27,540
3200	Autopsy table				
3220	Minimum	EA	660	19,370	20,030
3240	Maximum	"	660	27,370	28,030
3300	Incubators				
3320	15 cf	EA	330	9,670	10,000
3330	29 cf	"	550	13,110	13,660
3340	Infant transport, portable	"	350	7,280	7,630
3460	Headwall				
3465	Aluminum, with back frame and console	EA	330	5,460	5,790
6000	Hospital ground detection system				
6010	Power ground module	EA	190	1,710	1,900
6020	Ground slave module	"	140	720	860
6030	Master ground module	"	130	640	770

		UNIT	LABOR	MAT.	TOTAL
11700.10	**MEDICAL EQUIPMENT, Cont'd...**				
6040	Remote indicator	EA	130	670	800
6050	X-ray indicator	"	140	1,910	2,050
6060	Micro ammeter	"	170	2,270	2,440
6070	Supervisory module	"	140	1,910	2,050
6080	Ground cords	"	24.50	180	205
6100	Hospital isolation monitors, 5 ma				
6110	120v	EA	290	3,620	3,910
6120	208v	"	290	3,620	3,910
6130	240v	"	290	3,920	4,210
6210	Digital clock-timers separate display	"	130	1,590	1,720
6220	One display	"	130	1,020	1,150
6230	Remote control	"	100	500	600
6240	Battery pack	"	100	120	220
6310	Surgical chronometer clock and 3 timers	"	210	2,980	3,190
6320	Auxiliary control	"	96.00	810	906

TABLE OF CONTENTS   PAGE

DIVISION # 12 FURNISHINGS

		UNIT	LABOR	MAT.	TOTAL
12302.10	**CASEWORK**				
0080	Kitchen base cabinet, standard, 24" deep, 35" high				
0100	12" wide	EA	65.00	220	285
0120	18" wide	"	65.00	260	325
0140	24" wide	"	72.00	330	402
0160	27" wide	"	72.00	370	442
0180	36" wide	"	82.00	450	532
0200	48" wide	"	82.00	540	622
0210	Drawer base, 24" deep, 35" high				
0220	15" wide	EA	65.00	280	345
0230	18" wide	"	65.00	300	365
0240	24" wide	"	72.00	480	552
0250	27" wide	"	72.00	550	622
0260	30" wide	"	72.00	640	712
0270	Sink-ready base cabinet				
0280	30" wide	EA	72.00	300	372
0290	36" wide	"	72.00	310	382
0300	42" wide	"	72.00	340	412
0310	60" wide	"	82.00	410	492
0500	Corner cabinet, 36" wide	"	82.00	560	642
4000	Wall cabinet, 12" deep, 12" high				
4020	30" wide	EA	65.00	280	345
4060	36" wide	"	65.00	300	365
4110	24" high				
4120	30" wide	EA	72.00	370	442
4140	36" wide	"	72.00	380	452
4150	30" high				
4160	12" wide	EA	82.00	210	292
4200	24" wide	"	82.00	260	342
4320	30" wide	"	93.00	350	443
4340	36" wide	"	93.00	360	453
4350	Corner cabinet, 30" high				
4360	24" wide	EA	110	390	500
4390	36" wide	"	110	510	620
5020	Wardrobe	"	160	1,040	1,200
6980	Vanity with top, laminated plastic				
7000	24" wide	EA	160	860	1,020
7040	36" wide	"	220	1,110	1,330
7060	48" wide	"	260	1,240	1,500
12390.10	**COUNTERTOPS**				
1020	Stainless steel, countertop, with backsplash	SF	16.25	260	276
2000	Acid-proof, kemrock surface	"	10.75	100	111
12500.10	**WINDOW TREATMENT**				
1000	Drapery tracks, wall or ceiling mounted				
1040	Basic traverse rod				
1080	50 to 90"	EA	32.50	57.00	89.50
1100	84 to 156"	"	36.25	76.00	112
1120	136 to 250"	"	36.25	110	146
1140	165 to 312"	"	40.75	170	211
1160	Traverse rod with stationary curtain rod				
1180	30 to 50"	EA	32.50	86.00	119
1200	50 to 90"	"	32.50	98.00	131

		UNIT	LABOR	MAT.	TOTAL
12500.10	**WINDOW TREATMENT, Cont'd...**				
1220	84 to 156"	EA	36.25	140	176
1240	136 to 250"	"	40.75	170	211
1260	Double traverse rod				
1280	30 to 50"	EA	32.50	100	133
1300	50 to 84"	"	32.50	130	163
1320	84 to 156"	"	36.25	140	176
1340	136 to 250"	"	40.75	170	211
12510.10	**BLINDS**				
0990	Venetian blinds				
1000	2" slats	SF	1.63	39.50	41.13
1020	1" slats	"	1.63	42.25	43.88

Design Cost Data™ **DCD**

TABLE OF CONTENTS   PAGE

		UNIT	LABOR	MAT.	TOTAL
13116.10	**VAULTS**				
1000	Floor safes				
1010	1.0 cf	EA	54.00	930	984
1020	1.3 cf	"	82.00	1,030	1,112
13121.10	**PRE-ENGINEERED BUILDINGS**				
1080	Pre-engineered metal building, 40'x100'				
1100	14' eave height	SF	5.67	9.33	15.00
1120	16' eave height	"	6.54	10.50	17.04
1140	20' eave height	"	8.51	12.00	20.51
1150	60'x100'				
1160	14' eave height	SF	5.67	11.75	17.42
1180	16' eave height	"	6.54	13.00	19.54
1190	20' eave height	"	8.51	14.50	23.01
1195	80'x100'				
1200	14' eave height	SF	5.67	9.03	14.70
1210	16' eave height	"	6.54	9.33	15.87
1220	20' eave height	"	8.51	10.50	19.01
1280	100'x100'				
1300	14' eave height	SF	5.67	8.81	14.48
1320	16' eave height	"	6.54	9.18	15.72
1340	20' eave height	"	8.51	10.00	18.51
1350	100'x150'				
1360	14' eave height	SF	5.67	7.85	13.52
1380	16' eave height	"	6.54	8.15	14.69
1400	20' eave height	"	8.51	8.74	17.25
1410	120'x150'				
1420	14' eave height	SF	5.67	8.30	13.97
1440	16' eave height	"	6.54	8.44	14.98
1460	20' eave height	"	8.51	8.81	17.32
1480	140'x150'				
1500	14' eave height	SF	5.67	7.85	13.52
1520	16' eave height	"	6.54	8.05	14.59
1540	20' eave height	"	8.51	8.74	17.25
1600	160'x200'				
1620	14' eave height	SF	5.67	6.05	11.72
1640	16' eave height	"	6.54	6.24	12.78
1680	20' eave height	"	8.51	6.60	15.11
1690	200'x200'				
1700	14' eave height	SF	5.67	5.21	10.88
1720	16' eave height	"	6.54	5.73	12.27
1740	20' eave height	"	8.51	6.10	14.61
5020	Hollow metal door and frame, 6' x 7'	EA			1,390
5030	Sectional steel overhead door, manually operated				
5040	8' x 8'	EA			2,280
5080	12' x 12'	"			3,040
5100	Roll-up steel door, manually operated				
5120	10' x 10'	EA			1,770
5140	12' x 12'	"			3,180
5160	For gravity ridge ventilator with birdscreen	"			760
5161	9" throat x 10'	"			830
5181	12" throat x 10'	"			1,010
5200	For 20" rotary vent with damper	"			380

DIVISION # 13 SPECIAL CONSTRUCTION

		UNIT	LABOR	MAT.	TOTAL
13121.10	**PRE-ENGINEERED BUILDINGS, Cont'd...**				
5220	For 4' x 3' fixed louver	EA			270
5240	For 4' x 3' aluminum sliding window	"			240
5260	For 3' x 9' fiberglass panels	"			180
8020	Liner panel, 26 ga, painted steel	SF	1.80	3.34	5.14
8040	Wall panel insulated, 26 ga. steel, foam core	"	1.80	10.50	12.30
8060	Roof panel, 26 ga. painted steel	"	1.02	3.16	4.18
8080	Plastic (skylight)	"	1.02	7.12	8.14
9000	Insulation, 3-1/2" thick blanket, R11	"	0.48	2.13	2.61
13152.10	**SWIMMING POOL EQUIPMENT**				
1100	Diving boards				
1110	14' long				
1120	Aluminum	EA	280	4,970	5,250
1140	Fiberglass	"	280	3,760	4,040
1500	Ladders, heavy duty				
1510	2 steps				
1520	Minimum	EA	100	1,180	1,280
1540	Maximum	"	100	1,840	1,940
1550	4 steps				
1560	Minimum	EA	130	1,260	1,390
1580	Maximum	"	130	2,020	2,150
1600	Lifeguard chair				
1620	Minimum	EA	510	3,340	3,850
1640	Maximum	"	510	5,180	5,690
1700	Lights, underwater				
1705	12 volt, with transformer, 100 watt				
1710	Incandescent	EA	130	250	380
1715	Halogen	"	130	210	340
1720	LED	"	130	670	800
1730	110 volt				
1740	Minimum	EA	130	1,130	1,260
1760	Maximum	"	130	2,720	2,850
1780	Ground fault interrupter for 110 volt, each light	"	42.50	240	283
2000	Pool cover				
2020	Reinforced polyethylene	SF	3.92	2.41	6.33
2030	Vinyl water tube				
2040	Minimum	SF	3.92	1.48	5.40
2060	Maximum	"	3.92	2.20	6.12
2100	Slides with water tube				
2120	Minimum	EA	430	1,180	1,610
2140	Maximum	"	430	25,210	25,640
13152.20	**SAUNAS**				
0010	Prefabricated, cedar siding, insulated panels, prehung door,				
0020	4'x8"x4'-8"x6'-6"	EA			6,500
0030	5'-8"x6'-8"x6'-6"	"			7,870
0040	6'-8"x6'-8"x6'-6"	"			9,160
0050	7'-8"x7'-8"x6'-6"	"			11,060
0060	7'-8"x9'-8"x6'-6"	"			14,570
13200.10	**STORAGE TANKS**				
0080	Oil storage tank, underground, single wall, no excv.				
0090	Steel				
1000	500 gals	EA	360	4,040	4,400

		UNIT	LABOR	MAT.	TOTAL
13200.10	**STORAGE TANKS, Cont'd...**				
1020	1,000 gals	EA	480	5,470	5,950
1980	Fiberglass, double wall				
2000	550 gals	EA	480	11,370	11,850
2020	1,000 gals	"	480	14,620	15,100
2520	Above ground				
2530	Steel, single wall				
2540	275 gals	EA	290	2,290	2,580
2560	500 gals	"	480	5,720	6,200
2570	1,000 gals	"	570	7,810	8,380
3020	Fill cap	"	72.00	140	212
3040	Vent cap	"	72.00	140	212
3100	Level indicator	"	72.00	220	292

TABLE OF CONTENTS   PAGE

14210.10	ELEVATORS	UNIT	LABOR	MAT.	TOTAL
0120	Passenger elevators, electric, geared				
0502	Based on a shaft of 6 stops and 6 openings				
0510	50 fpm, 2000 lb	EA	2,850	145,640	148,490
0520	100 fpm, 2000 lb	"	3,170	151,020	154,190
0525	150 fpm				
0530	2000 lb	EA	3,560	166,620	170,180
0550	3000 lb	"	4,070	209,870	213,940
0560	4000 lb	"	4,750	218,380	223,130
1002	Based on a shaft of 8 stops and 8 openings				
1010	300 fpm				
1020	3000 lb	EA	5,700	270,190	275,890
1040	3500 lb	"	5,700	274,450	280,150
1060	4000 lb	"	6,330	287,960	294,290
1080	5000 lb	"	6,790	319,960	326,750
1502	Hydraulic, based on a shaft of 3 stops, 3 openings				
1508	50 fpm				
1510	2000 lb	EA	2,380	101,250	103,630
1520	2500 lb	"	2,380	108,270	110,650
1530	3000 lb	"	2,480	114,260	116,740
1600	For each additional; 50 fpm add per stop, $3500				
1620	500 lb, add per stop, $3500				
1630	Opening, add, $4200				
1640	Stop, add per stop, $5300				
1660	Bonderized steel door, add per opening, $400				
1670	Colored aluminum door, add per opening, $1500				
1680	Stainless steel door, add per opening, $650				
1690	Cast bronze door, add per opening, $1200				
1730	Custom cab interior, add per cab, $5000				
2000	Small elevators, 4 to 6 passenger capacity				
2005	Electric, push				
2010	2 stops	EA	2,380	34,510	36,890
2020	3 stops	"	2,590	43,140	45,730
2030	4 stops	"	2,850	49,110	51,960
14300.10	**ESCALATORS**				
1000	Escalators				
1020	32" wide, floor to floor				
1040	12' high	EA	4,750	180,150	184,900
1050	15' high	"	5,700	196,830	202,530
1060	18' high	"	7,130	211,990	219,120
1070	22' high	"	9,500	209,770	219,270
1080	25' high	"	11,400	238,380	249,780
1085	48" wide				
1090	12' high	EA	4,910	200,710	205,620
1100	15' high	"	5,940	219,050	224,990
1120	18' high	"	7,500	235,310	242,810
1130	22' high	"	10,180	263,410	273,590
1140	25' high	"	11,400	281,260	292,660
14410.10	**PERSONNEL LIFTS**				
1000	Electrically operated, 1 or 2 person lift				
1001	With attached foot platforms				
1020	3 stops	EA			12,390

		UNIT	LABOR	MAT.	TOTAL
14410.10	**PERSONNEL LIFTS, Cont'd...**				
1040	5 stops	EA			19,320
1060	7 stops	"			22,520
2000	For each additional stop, add $1250				
3020	Residential stair climber, per story	EA	550	5,750	6,300
14450.10	**VEHICLE LIFTS**				
1020	Automotive hoist, one post, semi-hydraulic, 8,000 lb	EA	2,850	4,140	6,990
1040	Full hydraulic, 8,000 lb	"	2,850	4,270	7,120
1060	2 post, semi-hydraulic, 10,000 lb	"	4,070	4,460	8,530
1070	Full hydraulic				
1080	10,000 lb	EA	4,070	5,250	9,320
1100	13,000 lb	"	7,130	6,570	13,700
1120	18,500 lb	"	7,130	10,500	17,630
1140	24,000 lb	"	7,130	14,770	21,900
1160	26,000 lb	"	7,130	14,380	21,510
1170	Pneumatic hoist, fully hydraulic				
1180	11,000 lb	EA	9,500	7,090	16,590
1200	24,000 lb	"	9,500	12,800	22,300
14560.10	**CHUTES**				
1020	Linen chutes, stainless steel, with supports				
1030	18" dia.	LF	5.14	170	175
1040	24" dia.	"	5.53	210	216
1050	30" dia.	"	6.00	220	226
1060	Hopper	EA	48.00	2,680	2,728
1070	Skylight	"	72.00	1,630	1,702
14580.10	**PNEUMATIC SYSTEMS**				
6000	Air-lift conveyor, 115 Volt, single phase, 100' long, 3" carrier	EA	6,040	4,570	10,610
6010	4" carrier	"	6,040	5,680	11,720
6020	6" carrier	"	6,040	10,320	16,360
7000	Pneumatic tube system accessories				
7010	Couplings and hanging accessories for 3" carrier system	LF			7.82
7020	For 4" carrier system	"			11.00
7030	For 6" carrier system	"			15.00
7040	24" CL. Expanded 90° bend, heavy duty, 3"-6" carrier systems	EA			140
7050	36" CL. Expanded 90° bend, heavy duty, 3"-6" carrier systems	"			330
7060	48" CL. Expanded 45° bend, heavy duty, 3" carrier system	"			110
7070	4" carrier system	"			150
7080	6" carrier system	"			220
7090	48" CL. Expanded 90° bend, heavy duty, 3" carrier system	"			140
8000	4" carrier system	"			190
8010	6" carrier system	"			330
8020	Stainless steel body up-grade, 3" carrier system	"			460
8030	4" carrier system	"			460
8040	6" carrier system	"			660
14580.30	**DUMBWAITERS**				
0010	28' travel, extruded alum., 4 stops, 100 lbs. capacity	EA			6,760
0020	150 lbs. capacity	"			9,510
0030	200 lbs. capacity	"			13,650

Design Cost Data

DCD

TABLE OF CONTENTS   PAGE

DIVISION # 15 MECHANICAL

		UNIT	LABOR	MAT.	TOTAL
15120.10	**BACKFLOW PREVENTERS**				
0080	Backflow preventer, flanged, cast iron, with valves				
0100	3" pipe	EA	360	4,030	4,390
15410.06	**C.I. PIPE, BELOW GROUND**				
1010	No hub pipe				
1020	1-1/2" pipe	LF	3.57	8.66	12.23
1030	2" pipe	"	3.97	8.89	12.86
1120	3" pipe	"	4.47	12.25	16.72
1220	4" pipe	"	5.96	16.00	21.96
15410.10	**COPPER PIPE**				
0880	Type K copper				
0900	1/2"	LF	2.23	3.77	6.00
1000	3/4"	"	2.38	7.03	9.41
1020	1"	"	2.55	9.20	11.75
3000	DWV, copper				
3020	1-1/4"	LF	2.98	10.25	13.23
3030	1-1/2"	"	3.25	13.00	16.25
3040	2"	"	3.57	17.00	20.57
3070	3"	"	3.97	29.00	32.97
3080	4"	"	4.47	50.00	54.47
3090	6"	"	5.10	200	205
6080	Type L copper				
6090	1/4"	LF	2.10	1.52	3.62
6095	3/8"	"	2.10	2.33	4.43
6100	1/2"	"	2.23	2.71	4.94
6190	3/4"	"	2.38	4.33	6.71
6240	1"	"	2.55	6.50	9.05
6580	Type M copper				
6600	1/2"	LF	2.23	1.91	4.14
6620	3/4"	"	2.38	3.12	5.50
6630	1"	"	2.55	5.06	7.61
15410.11	**COPPER FITTINGS**				
0850	Slip coupling				
0860	1/4"	EA	23.75	0.78	24.53
0870	1/2"	"	28.50	1.31	29.81
0880	3/4"	"	35.75	2.74	38.49
0890	1"	"	39.75	5.82	45.57
2660	Street ells, copper				
2670	1/4"	EA	28.50	5.69	34.19
2680	3/8"	"	32.50	3.92	36.42
2690	1/2"	"	35.75	1.58	37.33
2700	3/4"	"	37.75	3.33	41.08
2710	1"	"	39.75	8.62	48.37
4190	DWV fittings, coupling with stop				
4210	1-1/2"	EA	44.75	6.01	50.76
4230	2"	"	47.75	8.32	56.07
4260	3"	"	60.00	16.00	76.00
4280	3" x 2"	"	60.00	36.75	96.75
4290	4"	"	72.00	51.00	123
4300	Slip coupling				
4310	1-1/2"	EA	44.75	9.34	54.09
4320	2"	"	47.75	11.00	58.75

DIVISION # 15 MECHANICAL

		UNIT	LABOR	MAT.	TOTAL
15410.11	**COPPER FITTINGS, Cont'd...**				
4330	3"	EA	60.00	20.25	80.25
4340	90 ells				
4350	1-1/2"	EA	44.75	11.50	56.25
4360	1-1/2" x 1-1/4"	"	44.75	31.25	76.00
4370	2"	"	47.75	20.75	68.50
4380	2" x 1-1/2"	"	47.75	42.00	89.75
4390	3"	"	60.00	55.00	115
4400	4"	"	72.00	180	252
4410	Street, 90 elbows				
4420	1-1/2"	EA	44.75	14.50	59.25
4430	2"	"	47.75	31.75	79.50
4440	3"	"	60.00	81.00	141
4450	4"	"	72.00	200	272
5410	No-hub adapters				
5420	1-1/2" x 2"	EA	44.75	26.00	70.75
5430	2"	"	47.75	24.50	72.25
5440	2" x 3"	"	47.75	56.00	104
5450	3"	"	60.00	49.25	109
5460	3" x 4"	"	60.00	100	160
5470	4"	"	72.00	110	182
15410.82	**GALVANIZED STEEL PIPE**				
1000	Galvanized pipe				
1020	1/2" pipe	LF	7.15	3.01	10.16
1040	3/4" pipe	"	8.94	3.92	12.86
1200	90 degree ell, 150 lb malleable iron, galvanized				
1210	1/2"	EA	14.25	2.02	16.27
1220	3/4"	"	17.75	2.68	20.43
1400	45 degree ell, 150 lb m.i., galv.				
1410	1/2"	EA	14.25	3.23	17.48
1420	3/4"	"	17.75	4.38	22.13
1520	Tees, straight, 150 lb m.i., galv.				
1530	1/2"	EA	17.75	2.68	20.43
1540	3/4"	"	20.50	4.47	24.97
1800	Couplings, straight, 150 lb m.i., galv.				
1810	1/2"	EA	14.25	2.48	16.73
1820	3/4"	"	16.00	2.98	18.98
15430.23	**CLEANOUTS**				
0980	Cleanout, wall				
1000	2"	EA	47.75	240	288
1020	3"	"	47.75	340	388
1040	4"	"	60.00	340	400
1050	Floor				
1060	2"	EA	60.00	220	280
1080	3"	"	60.00	290	350
1100	4"	"	72.00	300	372
15430.25	**HOSE BIBBS**				
0005	Hose bibb				
0010	1/2"	EA	23.75	10.50	34.25
0200	3/4"	"	23.75	11.00	34.75

		UNIT	LABOR	MAT.	TOTAL
15430.60	**VALVES**				
0780	Gate valve, 125 lb, bronze, soldered				
0800	1/2"	EA	17.75	34.50	52.25
1000	3/4"	"	17.75	41.25	59.00
1280	Check valve, bronze, soldered, 125 lb				
1300	1/2"	EA	17.75	65.00	82.75
1320	3/4"	"	17.75	81.00	98.75
1790	Globe valve, bronze, soldered, 125 lb				
1800	1/2"	EA	20.50	80.00	101
1810	3/4"	"	22.25	99.00	121
15430.65	**VACUUM BREAKERS**				
1000	Vacuum breaker, atmospheric, threaded connection				
1010	3/4"	EA	28.50	55.00	83.50
1018	Anti-siphon, brass				
1020	3/4"	EA	28.50	60.00	88.50
15430.68	**STRAINERS**				
0980	Strainer, Y pattern, 125 psi, cast iron body, threaded				
1000	3/4"	EA	25.50	13.75	39.25
1980	250 psi, brass body, threaded				
2000	3/4"	EA	28.50	36.00	64.50
2130	Cast iron body, threaded				
2140	3/4"	EA	28.50	21.00	49.50
15430.70	**DRAINS, ROOF & FLOOR**				
1020	Floor drain, cast iron, with cast iron top				
1030	2"	EA	60.00	180	240
1040	3"	"	60.00	180	240
1050	4"	"	60.00	390	450
1090	Roof drain, cast iron				
1100	2"	EA	60.00	280	340
1110	3"	"	60.00	290	350
1120	4"	"	60.00	370	430
15440.10	**BATHS**				
0980	Bathtub, 5' long				
1000	Minimum	EA	240	580	820
1020	Average	"	360	1,270	1,630
1040	Maximum	"	720	2,900	3,620
1050	6' long				
1060	Minimum	EA	240	650	890
1080	Average	"	360	1,330	1,690
1100	Maximum	"	720	3,760	4,480
1110	Square tub, whirlpool, 4'x4'				
1120	Minimum	EA	360	2,000	2,360
1140	Average	"	720	2,830	3,550
1160	Maximum	"	890	8,640	9,530
1170	5'x5'				
1180	Minimum	EA	360	2,000	2,360
1200	Average	"	720	2,830	3,550
1220	Maximum	"	890	8,800	9,690
1230	6'x6'				
1240	Minimum	EA	360	2,430	2,790
1260	Average	"	720	3,560	4,280
1280	Maximum	"	890	10,200	11,090

DIVISION # 15 MECHANICAL

		UNIT	LABOR	MAT.	TOTAL
15440.10	**BATHS, Cont'd...**				
8980	For trim and rough-in				
9000	Minimum	EA	240	210	450
9020	Average	"	360	300	660
9040	Maximum	"	720	860	1,580
15440.12	**DISPOSALS & ACCESSORIES**				
0040	Disposal, continuous feed				
0050	Minimum	EA	140	79.00	219
0060	Average	"	180	220	400
0070	Maximum	"	240	420	660
0200	Batch feed, 1/2 hp				
0220	Minimum	EA	140	300	440
0230	Average	"	180	600	780
0240	Maximum	"	240	1,040	1,280
1100	Hot water dispenser				
1110	Minimum	EA	140	220	360
1120	Average	"	180	350	530
1130	Maximum	"	240	560	800
1140	Epoxy finish faucet	"	140	310	450
1160	Lock stop assembly	"	89.00	67.00	156
1170	Mounting gasket	"	60.00	7.74	67.74
1180	Tailpipe gasket	"	60.00	1.13	61.13
1190	Stopper assembly	"	72.00	26.50	98.50
1200	Switch assembly, on/off	"	120	30.25	150
1210	Tailpipe gasket washer	"	35.75	1.21	36.96
1220	Stop gasket	"	39.75	2.66	42.41
1230	Tailpipe flange	"	35.75	0.30	36.05
1240	Tailpipe	"	44.75	3.44	48.19
15440.15	**FAUCETS**				
0980	Kitchen				
1000	Minimum	EA	120	91.00	211
1020	Average	"	140	250	390
1040	Maximum	"	180	310	490
1050	Bath				
1060	Minimum	EA	120	91.00	211
1080	Average	"	140	270	410
1100	Maximum	"	180	410	590
1110	Lavatory, domestic				
1120	Minimum	EA	120	97.00	217
1140	Average	"	140	310	450
1160	Maximum	"	180	510	690
1290	Washroom				
1300	Minimum	EA	120	120	240
1320	Average	"	140	300	440
1340	Maximum	"	180	560	740
1350	Handicapped				
1360	Minimum	EA	140	130	270
1380	Average	"	180	400	580
1400	Maximum	"	240	620	860
1410	Shower				
1420	Minimum	EA	120	120	240
1440	Average	"	140	350	490

		UNIT	LABOR	MAT.	TOTAL
15440.15	**FAUCETS, Cont'd...**				
1460	Maximum	EA	180	560	740
1480	For trim and rough-in				
1500	Minimum	EA	140	85.00	225
1520	Average	"	180	130	310
1540	Maximum	"	360	220	580
15440.18	**HYDRANTS**				
0980	Wall hydrant				
1000	8" thick	EA	120	400	520
1020	12" thick	"	140	470	610
15440.20	**LAVATORIES**				
1980	Lavatory, countertop, porcelain enamel on cast iron				
2000	Minimum	EA	140	210	350
2010	Average	"	180	320	500
2020	Maximum	"	240	570	810
2080	Wall hung, china				
2100	Minimum	EA	140	290	430
2110	Average	"	180	340	520
2120	Maximum	"	240	850	1,090
2280	Handicapped				
2300	Minimum	EA	180	470	650
2310	Average	"	240	540	780
2320	Maximum	"	360	910	1,270
8980	For trim and rough-in				
9000	Minimum	EA	180	240	420
9020	Average	"	240	410	650
9040	Maximum	"	360	510	870
15440.30	**SHOWERS**				
0980	Shower, fiberglass, 36"x34"x84"				
1000	Minimum	EA	510	630	1,140
1020	Average	"	720	880	1,600
1040	Maximum	"	720	1,270	1,990
2980	Steel, 1 piece, 36"x36"				
3000	Minimum	EA	510	580	1,090
3020	Average	"	720	880	1,600
3040	Maximum	"	720	1,040	1,760
3980	Receptor, molded stone, 36"x36"				
4000	Minimum	EA	240	240	480
4020	Average	"	360	410	770
4040	Maximum	"	600	630	1,230
8980	For trim and rough-in				
9000	Minimum	EA	330	240	570
9020	Average	"	400	410	810
9040	Maximum	"	720	510	1,230
15440.40	**SINKS**				
0980	Service sink, 24"x29"				
1000	Minimum	EA	180	700	880
1020	Average	"	240	870	1,110
1040	Maximum	"	360	1,280	1,640
2000	Kitchen sink, single, stainless steel, single bowl				
2020	Minimum	EA	140	310	450
2040	Average	"	180	350	530

		UNIT	LABOR	MAT.	TOTAL
15440.40	**SINKS, Cont'd...**				
2060	Maximum	EA	240	640	880
2070	Double bowl				
2080	Minimum	EA	180	350	530
2100	Average	"	240	390	630
2120	Maximum	"	360	680	1,040
2190	Porcelain enamel, cast iron, single bowl				
2200	Minimum	EA	140	220	360
2220	Average	"	180	290	470
2240	Maximum	"	240	450	690
2250	Double bowl				
2260	Minimum	EA	180	300	480
2280	Average	"	240	420	660
2300	Maximum	"	360	600	960
2980	Mop sink, 24"x36"x10"				
3000	Minimum	EA	140	530	670
3020	Average	"	180	640	820
3040	Maximum	"	240	860	1,100
5980	Washing machine box				
6000	Minimum	EA	180	190	370
6040	Average	"	240	280	520
6060	Maximum	"	360	340	700
8980	For trim and rough-in				
9000	Minimum	EA	240	310	550
9020	Average	"	360	480	840
9040	Maximum	"	480	620	1,100
15440.60	**WATER CLOSETS**				
0980	Water closet flush tank, floor mounted				
1000	Minimum	EA	180	360	540
1010	Average	"	240	710	950
1020	Maximum	"	360	1,130	1,490
1030	Handicapped				
1040	Minimum	EA	240	490	730
1050	Average	"	360	890	1,250
1060	Maximum	"	720	1,680	2,400
8980	For trim and rough-in				
9000	Minimum	EA	180	230	410
9020	Average	"	240	270	510
9040	Maximum	"	360	360	720
15440.70	**DOMESTIC WATER HEATERS**				
0980	Water heater, electric				
1000	6 gal	EA	120	450	570
1020	10 gal	"	120	460	580
1030	15 gal	"	120	450	570
1040	20 gal	"	140	630	770
1050	30 gal	"	140	650	790
1060	40 gal	"	140	710	850
1070	52 gal	"	180	800	980
2980	Oil fired				
3000	20 gal	EA	360	1,430	1,790
3020	50 gal	"	510	2,230	2,740

DIVISION # 15 MECHANICAL

		UNIT	LABOR	MAT.	TOTAL
15610.10	**FURNACES**				
0980	Electric, hot air				
1000	40 mbh	EA	360	850	1,210
1020	60 mbh	"	380	920	1,300
1040	80 mbh	"	400	1,000	1,400
1060	100 mbh	"	420	1,130	1,550
1080	125 mbh	"	430	1,380	1,810
1980	Gas fired hot air				
2000	40 mbh	EA	360	850	1,210
2020	60 mbh	"	380	910	1,290
2040	80 mbh	"	400	1,050	1,450
2060	100 mbh	"	420	1,090	1,510
2080	125 mbh	"	430	1,200	1,630
2980	Oil fired hot air				
3000	40 mbh	EA	360	1,140	1,500
3020	60 mbh	"	380	1,890	2,270
3040	80 mbh	"	400	1,900	2,300
3060	100 mbh	"	420	1,930	2,350
3080	125 mbh	"	430	2,000	2,430
15780.20	**ROOFTOP UNITS**				
0980	Packaged, single zone rooftop unit, with roof curb				
1000	2 ton	EA	720	4,130	4,850
1020	3 ton	"	720	4,340	5,060
1040	4 ton	"	890	4,740	5,630
15830.70	**UNIT HEATERS**				
0980	Steam unit heater, horizontal				
1000	12,500 btuh, 200 cfm	EA	120	560	680
1010	17,000 btuh, 300 cfm	"	120	740	860
15855.10	**AIR HANDLING UNITS**				
0980	Air handling unit, medium pressure, single zone				
1000	1500 cfm	EA	450	4,840	5,290
1060	3000 cfm	"	790	6,360	7,150
8980	Rooftop air handling units				
9000	4950 cfm	EA	790	13,910	14,700
9060	7370 cfm	"	1,020	17,640	18,660
15870.20	**EXHAUST FANS**				
0160	Belt drive roof exhaust fans				
1020	640 cfm, 2618 fpm	EA	89.00	1,140	1,229
1030	940 cfm, 2604 fpm	"	89.00	1,480	1,569
15890.10	**METAL DUCTWORK**				
0090	Rectangular duct				
0100	Galvanized steel				
1000	Minimum	LB	6.50	0.88	7.38
1010	Average	"	7.94	1.10	9.04
1020	Maximum	"	12.00	1.68	13.68
1080	Aluminum				
1100	Minimum	LB	14.25	2.29	16.54
1120	Average	"	17.75	3.05	20.80
1140	Maximum	"	23.75	3.79	27.54
1160	Fittings				
1180	Minimum	EA	23.75	7.26	31.01
1200	Average	"	35.75	11.00	46.75

DIVISION # 15 MECHANICAL

		UNIT	LABOR	MAT.	TOTAL
15890.10	**METAL DUCTWORK, Cont'd...**				
1220	Maximum	EA	72.00	16.00	88.00
15890.30	**FLEXIBLE DUCTWORK**				
1010	Flexible duct, 1.25" fiberglass				
1020	5" dia.	LF	3.57	3.31	6.88
1040	6" dia.	"	3.97	3.68	7.65
1060	7" dia.	"	4.20	4.54	8.74
1080	8" dia.	"	4.47	4.76	9.23
1100	10" dia.	"	5.10	6.34	11.44
1120	12" dia.	"	5.50	6.93	12.43
9000	Flexible duct connector, 3" wide fabric	"	12.00	2.31	14.31
15910.10	**DAMPERS**				
0980	Horizontal parallel aluminum backdraft damper				
1000	12" x 12"	EA	17.75	58.00	75.75
1010	16" x 16"	"	20.50	60.00	80.50
15940.10	**DIFFUSERS**				
1980	Ceiling diffusers, round, baked enamel finish				
2000	6" dia.	EA	23.75	40.25	64.00
2020	8" dia.	"	29.75	48.50	78.25
2040	10" dia.	"	29.75	54.00	83.75
2060	12" dia.	"	29.75	69.00	98.75
2480	Rectangular				
2500	6x6"	EA	23.75	43.00	66.75
2520	9x9"	"	35.75	52.00	87.75
2540	12x12"	"	35.75	76.00	112
2560	15x15"	"	35.75	95.00	131
2580	18x18"	"	35.75	120	156

Design Cost Data™

DCD

TABLE OF CONTENTS   PAGE

		UNIT	LABOR	MAT.	TOTAL
16050.30	**BUS DUCT**				
1000	Bus duct, 100a, plug-in				
1010	10', 600v	EA	230	320	550
1020	With ground	"	350	430	780
1145	Circuit breakers, with enclosure				
1147	1 pole				
1150	15a-60a	EA	83.00	320	403
1160	70a-100a	"	100	370	470
1165	2 pole				
1170	15a-60a	EA	91.00	450	541
1180	70a-100a	"	110	540	650
16110.22	**EMT CONDUIT**				
0080	EMT conduit				
0100	1/2"	LF	2.50	0.60	3.10
1020	3/4"	"	3.32	1.09	4.41
1030	1"	"	4.15	1.82	5.97
2980	90 deg. elbow				
3000	1/2"	EA	7.37	5.63	13.00
3040	3/4"	"	8.30	6.19	14.49
3060	1"	"	8.85	9.55	18.40
3980	Connector, steel compression				
4000	1/2"	EA	7.37	1.72	9.09
4040	3/4"	"	7.37	3.30	10.67
4060	1"	"	7.37	4.97	12.34
0080	Flexible conduit, steel				
0100	3/8"	LF	2.50	0.75	3.25
1020	1/2	"	2.50	0.85	3.35
1040	3/4"	"	3.32	1.17	4.49
1060	1"	"	3.32	2.23	5.55
16110.24	**GALVANIZED CONDUIT**				
1980	Galvanized rigid steel conduit				
2000	1/2"	LF	3.32	2.82	6.14
2040	3/4"	"	4.15	3.13	7.28
2060	1"	"	4.91	4.51	9.42
2080	1-1/4"	"	6.64	6.24	12.88
2100	1-1/2"	"	7.37	7.34	14.71
2120	2"	"	8.30	9.34	17.64
2480	90 degree ell				
2500	1/2"	EA	20.75	7.51	28.26
2540	3/4"	"	25.50	7.85	33.35
2560	1"	"	31.50	12.00	43.50
2580	1-1/4"	"	37.00	16.50	53.50
2590	1-1/2"	"	41.50	20.50	62.00
2600	2"	"	44.25	29.50	73.75
3200	Couplings, with set screws				
3220	1/2"	EA	4.15	3.76	7.91
3260	3/4"	"	4.91	4.96	9.87
3280	1"	"	6.64	7.90	14.54
3300	1-1/4"	"	8.30	13.50	21.80
3320	1-1/2"	"	10.25	17.50	27.75
3340	2"	"	12.00	39.25	51.25

		UNIT	LABOR	MAT.	TOTAL
16110.25	**PLASTIC CONDUIT**				
3010	PVC conduit, schedule 40				
3020	1/2"	LF	2.50	0.73	3.23
3040	3/4"	"	2.50	0.91	3.41
3060	1"	"	3.32	1.32	4.64
3080	1-1/4"	"	3.32	1.82	5.14
3100	1-1/2"	"	4.15	2.17	6.32
3120	2"	"	4.15	2.78	6.93
3480	Couplings				
3500	1/2"	EA	4.15	0.45	4.60
3520	3/4"	"	4.15	0.53	4.68
3540	1"	"	4.15	0.84	4.99
3560	1-1/4"	"	4.91	1.11	6.02
3580	1-1/2"	"	4.91	1.54	6.45
3600	2"	"	4.91	2.02	6.93
3705	90 degree elbows				
3710	1/2"	EA	8.30	1.74	10.04
3740	3/4"	"	10.25	1.90	12.15
3760	1"	"	10.25	3.01	13.26
3780	1-1/4"	"	12.00	4.20	16.20
3800	1-1/2"	"	15.75	5.69	21.44
3810	2"	"	18.50	7.94	26.44
16110.28	**STEEL CONDUIT**				
7980	Intermediate metal conduit (IMC)				
8000	1/2"	LF	2.50	2.02	4.52
8040	3/4"	"	3.32	2.48	5.80
8060	1"	"	4.15	3.76	7.91
8080	1-1/4"	"	4.91	4.81	9.72
8100	1-1/2"	"	6.64	6.02	12.66
8120	2"	"	7.37	7.86	15.23
8490	90 degree ell				
8500	1/2"	EA	20.75	15.50	36.25
8540	3/4"	"	25.50	16.25	41.75
8560	1"	"	31.50	24.75	56.25
8580	1-1/4"	"	37.00	34.50	71.50
8600	1-1/2"	"	41.50	42.50	84.00
8620	2"	"	47.50	61.00	109
9260	Couplings				
9280	1/2"	EA	4.15	3.79	7.94
9290	3/4"	"	4.91	4.66	9.57
9300	1"	"	6.64	6.90	13.54
9310	1-1/4"	"	7.37	8.64	16.01
9320	1-1/2"	"	8.30	11.00	19.30
9330	2"	"	8.85	14.50	23.35
16110.35	**SURFACE MOUNTED RACEWAY**				
0980	Single Raceway				
1000	3/4" x 17/32" Conduit	LF	3.32	1.83	5.15
1020	Mounting Strap	EA	4.42	0.49	4.91
1040	Connector	"	4.42	0.66	5.08
1060	Elbow				
2000	45 degree	EA	4.15	8.38	12.53
2020	90 degree	"	4.15	2.67	6.82

		UNIT	LABOR	MAT.	TOTAL
16110.35	**SURFACE MOUNTED RACEWAY, Cont'd...**				
2040	internal	EA	4.15	3.36	7.51
2050	external	"	4.15	3.10	7.25
2060	Switch	"	33.25	21.75	55.00
2100	Utility Box	"	33.25	14.50	47.75
2110	Receptacle	"	33.25	25.75	59.00
2140	3/4" x 21/32" Conduit	LF	3.32	2.09	5.41
2160	Mounting Strap	EA	4.42	0.77	5.19
2180	Connector	"	4.42	0.79	5.21
2200	Elbow				
2210	45 degree	EA	4.15	10.25	14.40
2220	90 degree	"	4.15	2.85	7.00
2240	internal	"	4.15	3.87	8.02
2260	external	"	4.15	3.87	8.02
3000	Switch	"	33.25	21.75	55.00
3010	Utility Box	"	33.25	14.50	47.75
3020	Receptacle	"	33.25	25.75	59.00
16120.43	**COPPER CONDUCTORS**				
0980	Copper conductors, type THW, solid				
1000	#14	LF	0.33	0.12	0.45
1040	#12	"	0.41	0.18	0.59
1060	#10	"	0.50	0.28	0.78
2010	THHN-THWN, solid				
2020	#14	LF	0.33	0.12	0.45
2040	#12	"	0.41	0.18	0.59
2060	#10	"	0.50	0.28	0.78
6215	Type BX solid armored cable				
6220	#14/2	LF	2.07	0.81	2.88
6230	#14/3	"	2.32	1.28	3.60
6240	#14/4	"	2.55	1.80	4.35
6250	#12/2	"	2.32	0.83	3.15
6260	#12/3	"	2.55	1.34	3.89
6270	#12/4	"	2.88	1.85	4.73
6280	#10/2	"	2.55	1.55	4.10
6290	#10/3	"	2.88	2.22	5.10
6300	#10/4	"	3.32	3.45	6.77
16120.47	**SHEATHED CABLE**				
6700	Non-metallic sheathed cable				
6705	Type NM cable with ground				
6710	#14/2	LF	1.24	0.28	1.52
6720	#12/2	"	1.32	0.44	1.76
6730	#10/2	"	1.47	0.69	2.16
6740	#8/2	"	1.66	1.13	2.79
6750	#6/2	"	2.07	1.78	3.85
6760	#14/3	"	2.14	0.39	2.53
6770	#12/3	"	2.21	0.62	2.83
6780	#10/3	"	2.25	0.99	3.24
6790	#8/3	"	2.28	1.66	3.94
6800	#6/3	"	2.32	2.68	5.00
6810	#4/3	"	2.65	5.55	8.20
6820	#2/3	"	2.88	8.34	11.22

		UNIT	LABOR	MAT.	TOTAL
16130.40	**BOXES**				
5000	Round cast box, type SEH				
5010	1/2"	EA	28.75	20.75	49.50
5020	3/4"	"	35.00	20.75	55.75
16130.60	**PULL AND JUNCTION BOXES**				
1050	4"				
1060	Octagon box	EA	9.48	3.96	13.44
1070	Box extension	"	4.91	6.67	11.58
1080	Plaster ring	"	4.91	3.66	8.57
1100	Cover blank	"	4.91	1.61	6.52
1120	Square box	"	9.48	5.70	15.18
1140	Box extension	"	4.91	5.59	10.50
1160	Plaster ring	"	4.91	3.06	7.97
1180	Cover blank	"	4.91	1.57	6.48
16130.80	**RECEPTACLES**				
0500	Contractor grade duplex receptacles, 15a 120v				
0510	Duplex	EA	16.50	1.60	18.10
1000	125 volt, 20a, duplex, standard grade	"	16.50	12.00	28.50
1040	Ground fault interrupter type	"	24.50	38.75	63.25
1520	250 volt, 20a, 2 pole, single, grounding type	"	16.50	20.00	36.50
1540	120/208v, 4 pole, single receptacle, twist lock				
1560	20a	EA	28.75	23.75	52.50
1580	50a	"	28.75	45.25	74.00
1590	125/250v, 3 pole, flush receptacle				
1600	30a	EA	24.50	24.00	48.50
1620	50a	"	24.50	29.75	54.25
1640	60a	"	28.75	77.00	106
16350.10	**CIRCUIT BREAKERS**				
5000	Load center circuit breakers, 240v				
5010	1 pole, 10-60a	EA	20.75	21.25	42.00
5015	2 pole				
5020	10-60a	EA	33.25	43.25	76.50
5030	70-100a	"	55.00	130	185
5040	110-150a	"	60.00	280	340
5065	Load center, GFI breakers, 240v				
5070	1 pole, 15-30a	EA	24.50	160	185
5095	Tandem breakers, 240v				
5100	1 pole, 15-30a	EA	33.25	35.00	68.25
5110	2 pole, 15-30a	"	44.25	64.00	108
16395.10	**GROUNDING**				
0500	Ground rods, copper clad, 1/2" x				
0510	6'	EA	55.00	17.00	72.00
0520	8'	"	60.00	23.50	83.50
0535	5/8" x				
0550	6'	EA	60.00	22.75	82.75
0560	8'	"	83.00	29.25	112
16430.20	**METERING**				
0500	Outdoor wp meter sockets, 1 gang, 240v, 1 phase				
0510	Includes sealing ring, 100a	EA	130	47.25	177
0520	150a	"	150	56.00	206
0530	200a	"	170	71.00	241

DIVISION # 16 ELECTRICAL

		UNIT	LABOR	MAT.	TOTAL
16470.10	**PANELBOARDS**				
1000	Indoor load center, 1 phase 240v main lug only				
1020	30a - 2 spaces	EA	170	32.25	202
1030	100a - 8 spaces	"	200	100	300
1040	150a - 16 spaces	"	250	270	520
1050	200a - 24 spaces	"	290	550	840
1060	200a - 42 spaces	"	330	570	900
16490.10	**SWITCHES**				
4000	Photoelectric switches				
4010	1000 watt				
4020	105-135v	EA	60.00	37.00	97.00
4970	Dimmer switch and switch plate				
4990	600w	EA	25.50	34.00	59.50
5171	Contractor grade wall switch 15a, 120v				
5172	Single pole	EA	13.25	1.79	15.04
5173	Three way	"	16.50	3.26	19.76
5174	Four way	"	22.25	11.00	33.25
16510.05	**INTERIOR LIGHTING**				
0005	Recessed fluorescent fixtures, 2'x2'				
0010	2 lamp	EA	60.00	72.00	132
0020	4 lamp	"	60.00	98.00	158
0205	Surface mounted incandescent fixtures				
0210	40w	EA	55.00	110	165
0220	75w	"	55.00	110	165
0230	100w	"	55.00	120	175
0240	150w	"	55.00	160	215
0289	Recessed incandescent fixtures				
0290	40w	EA	130	150	280
0300	75w	"	130	160	290
0310	100w	"	130	180	310
0320	150w	"	130	190	320
0395	Light track single circuit				
0400	2'	EA	41.50	44.25	85.75
0410	4'	"	41.50	52.00	93.50
0420	8'	"	83.00	72.00	155
0430	12'	"	130	100	230
16510.10	**INDUSTRIAL LIGHTING**				
0500	Strip fluorescent				
0510	4'				
0520	1 lamp	EA	55.00	45.50	101
0540	2 lamps	"	55.00	55.00	110
0550	8'				
0560	1 lamp	EA	60.00	66.00	126
0580	2 lamps	"	74.00	99.00	173
1000	Parabolic troffer, 2'x2'				
1020	With 2 U lamps	EA	83.00	130	213
1060	With 3 U lamps	"	95.00	150	245
1080	2'x4'				
1100	With 2 40w lamps	EA	95.00	150	245
1120	With 3 40w lamps	"	110	150	260
1140	With 4 40w lamps	"	110	160	270
3120	High pressure sodium, hi-bay open				

16 - 6

© 2020 By Design & Construction Resources

		UNIT	LABOR	MAT.	TOTAL
16510.10	**INDUSTRIAL LIGHTING, Cont'd...**				
3140	400w	EA	140	460	600
3160	1000w	"	200	790	990
3170	Enclosed				
3180	400w	EA	200	740	940
3200	1000w	"	250	1,030	1,280
3210	Metal halide hi-bay, open				
3220	400w	EA	140	280	420
3240	1000w	"	200	580	780
3250	Enclosed				
3260	400w	EA	200	640	840
3280	1000w	"	250	610	860
3590	Metal halide, low-bay, pendant mounted				
3600	175w	EA	110	370	480
3620	250w	"	130	510	640
3660	400w	"	180	550	730
16710.10	**COMMUNICATIONS COMPONENTS**				
0020	Port desktop switch unit-4	EA			85.00
0030	(USB)-4	"			220
0040	8	"			360
0050	16	"			600
0060	32	"			3,870
0070	Port console unit-8	"			2,420
0080	16	"			3,030
0090	Cat 5EJack and RJ45 coupler	"			4.11
1000	Quick-Port and voice grade	"			4.65
1010	Trac-Jack category 5E and connectors	"			4.63
1020	Snap-In connector	"			6.35
1030	Fast ethernet media converter	"			150
1040	Gigabit media converter	"			420
1050	With link fault signaling	"			150
1060	Gigabit switching media converter	"			600
1070	16-Bay media chassis	"			740
1080	Fiber-optic cable, Single-Mode, 50/125 microns, 10' length	"			24.25
1090	Multi-Mode, 62.5/125 microns, 10' length	"			27.75
2000	Fiber-Optic connectors, Unicam	"			18.75
2010	Fast-cam	"			19.25
2020	Adhesive style	"			6.95
2030	Threat-lock	"			12.00
2040	Unicam, high performance, single-mode	"			24.25
2050	multi-mode	"			21.75
2060	Network Cable, Cat5, solid PVC, 50'	"			17.50
2070	1000' Cat5, stranded PVC	"			200
2080	plenum PVC	"			240
2090	1000' Cat6, solid PVC	"			170
3000	USB Cables, 5 in 1 connector, male and female	"			21.75
3010	3 in 1 quick-connect, 4-pin or 6-pin	"			26.50
3020	Squid hub	"			36.25
16760.10	**AUDIO/VIDEO COMPONENTS**				
0010	Keystone jack module	EA			3.30
0020	F-Type quick port	"			2.31
0030	BNC quick port	"			5.50

		UNIT	LABOR	MAT.	TOTAL
16760.10	**AUDIO/VIDEO COMPONENTS, Cont'd...**				
0040	RCA jack bulkhead connector	EA			4.40
0050	Quick-Port	"			5.72
0060	Snap-in	"			8.80
0070	HDMI wall plate and jack	"			10.50
0080	Voice/Data adapter	"			3.85
16760.20	**AUDIO/VIDEO CABLES**				
0010	Monster cable, 24k gold plated	EA			26.00
0020	DVI-HDMI	"			11.25
0030	Satellite/Video	"			3.76
0040	Digital/Optical	"			12.00
0050	Video/RCA	"			7.05
0060	S-Video	"			7.84
0070	Monster/S-video	"			41.75
0080	Speaker, clear jacket	FT			0.89
0090	Speaker, high performance	"			2.09
0100	Flex-Premiere, oxygen free	"			0.31
16850.10	**ELECTRIC HEATING**				
1000	Baseboard heater				
1020	2', 375w	EA	83.00	41.75	125
1040	3', 500w	"	83.00	49.50	133
1060	4', 750w	"	95.00	55.00	150
1100	5', 935w	"	110	78.00	188
1120	6', 1125w	"	130	92.00	222
1140	7', 1310w	"	150	100	250
1160	8', 1500w	"	170	120	290
1180	9', 1680w	"	180	130	310
1200	10', 1875w	"	190	180	370
1210	Unit heater, wall mounted				
1215	750w	EA	130	160	290
1220	1500w	"	140	220	360
16910.40	**CONTROL CABLE**				
0980	Control cable, 600v, #14 THWN, PVC jacket				
1000	2 wire	LF	0.66	0.38	1.04
1020	4 wire	"	0.83	0.65	1.48

Square Foot Tables

The following Square Foot Tables list hundreds of actual projects for ten building types, each with associated building size, total square foot building cost and percentage of project costs for total mechanical and electrical components. This data provides an overview of construction costs by building type. These costs are for actual projects. The variations within similar building types may be due, among other factors, to size, location, quality and specified components, materials and processes. Depending upon all such factors, specific building costs can vary significantly and may not necessarily fall within the range of costs as presented. The data has been updated to reflect current construction costs.

All prices are updated to January 1, 2021 and are national averages.
For a more in-depth report of any of these buildings or additional case studies, contact
Design Cost Data at 800-533-5680, or go to www.DCD.com

SF-1

Design Cost Data™

DCD

2021 SQUARE FOOT TABLES

PROJECT	DESCRIPTION	CITY	STATE	SIZE	$/SF	NOTES
		Commercial				
Bank	School Credit Union Administration Building	Katy	TX	30,700	$216.78	New
	FineMark National Bank & Trust	Fort Myers	FL	20,039	$408.96	New
	Florida Shores Bank	Pompano Beach	FL	11,697	$520.52	New
	Mobiloil Credit Union	Vidor	TX	9,252	$365.04	New
	Beaumont Community Credit Union	Beaumont	TX	3,267	$413.83	New
Office	Allendale Town Center	Allendale	NJ	80,226	$38.20	Addition/Renovation
	Roanoke Electric Cooperative	Ahoskie	NC	52,752	$217.92	New
	Regional Aviation & Training Center	Currituck	NC	39,930	$276.44	New
	Transportation/Warehouse Facility	Monroe	GA	32,400	$191.90	New
	Collection System Operations Facility	Walnut Creek	CA	27,179	$379.11	New
	ULTA - (Shell Only)	Pensacola	FL	10,850	$98.09	Renovation
	Daycare Center	Clawson	MI	4,270	$155.32	Adaptive Reuse
	Campgrounds Office & Retail	Cincinnati	OH	2,300	$340.27	New
Parking	Palm Avenue Parking Garage	Sarasota	FL	287,040	$58.13	New
	Reynolds Street Parking Deck	Augusta	GA	214,000	$69.15	New
	Awty Int. School Parking Structure	Houston	TX	174,582	$61.22	New
Retail	SNG Center - Mixed-Use	Fargo	ND	143,860	$102.89	New
	Roof & Lifeway/Steinmart Renovation	Pensacola	FL	88,299	$36.14	Renovation
	No Frills Supermarket	Omaha	NE	61,000	$110.07	New
	West Oaks Mall Redevelopment	Houston	TX	49,800	$195.89	Renovation
	World of Decor	Deerfield Beach	FL	47,500	$149.51	New
	Sarasota Yacht Club	Sarasota	FL	41,332	$439.28	New
	Karschs Village Market	Barnhart	MO	35,384	$85.80	New
	Fresh Thyme Farmers Market	Fishers	IN	28,784	$188.13	New
	Nashville Hangar Inc.	Nashville	TN	28,702	$197.26	New
	Party Time Plus	Billings	MT	26,000	$100.08	New
	Marshalls - (Shell Only)	Pensacola	FL	25,990	$47.89	Renovation
	Ed Hicks Mercedes-Benz USA	Corpus Christi	TX	25,273	$261.77	New
	Montana Honda & Marine	Billings	MT	22,963	$113.75	Addition
	Fresh Market - (Shell Only)	Pensacola	FL	21,000	$82.32	Renovation
	Theatre Exchange Interior Fit Up	Manitoba	CA	19,344	$132.78	Renovation
	DSW Shoes Renovation	Pensacola	FL	18,000	$92.17	Renovation
	Dormans Lighting & Design	Lutherville	MD	15,220	$124.94	Addition
	The Groves Exterior Renovation	Farmington	MI	15,137	$70.63	Renovation
	Don Gibson Theatre	Shelby	NC	13,386	$372.40	Renovation
	Tri Ford Showroom Expansion	Highland	IL	12,881	$121.80	Addition/Renovation
	Fiat of LeHigh Valley	Easton	PA	11,905	$155.97	New
	Sicardi Art Gallery	Houston	TX	6,175	$245.35	New
	Childrens Mercy Hospital Gift Shop	Kansas City	MO	5,010	$309.23	Tenant Build-out
Restaurant	LaMar Cebicheria Peruana Restaurant	San Francisco	CA	11,000	$316.16	Tenant Build-out
	Ulele Restaurant	Tampa	FL	8,905	$766.28	Adaptive Reuse
	Youells Oyster House	Allentown	PA	6,107	$251.55	New
	Mellow Mushroom Highlands Shell	Louisville	KY	5,802	$101.96	New
	Mellow Mushroom Highlands TBO	Louisville	KY	5,802	$169.11	Tenant Build-out
	Mellow Mushroom Pizza	Wilder	KY	5,500	$241.87	New
	Liberty Microbrewery	Plymouth	MI	3,425	$174.02	Addition
	700 South Deli	Linthicum	MD	3,200	$214.42	Tenant Build-out
	New York Pizza Department (NYPD)	Tempe	AZ	2,338	$328.05	Tenant Build-out
	Airport Restaurant Build Out	Eglin Air Force Base	FL	2,320	$282.41	Tenant Build-out

All prices are updated to January 1, 2021 and are national averages.
For a more in-depth report of any of these buildings or additional case studies, contact
Design, Cost and Data at 800-533-5680, or go to www.DCD.com

2021 SQUARE FOOT TABLES

PROJECT	DESCRIPTION	CITY	STATE	SIZE	$/SF	NOTES
		Civic/Government				
Civic Center	Lincoln Center	Fort Collins	CO	38,160	$216.29	Addition/Renovation
	Rockport City Services Building	Rockport	TX	20,062	$223.75	New
	Sinclair Park Community Centre	Manitoba	CA	17,007	$316.81	Addition/Renovation
	Mt. Olive City Hall Complex	Mount Olive	IL	14,360	$121.48	New
	Teaneck Municipal Complex	Teaneck	NJ	12,870	$232.81	Addition/Renovation
	Cobb Community Center Additions	Pensacola	FL	5,200	$324.01	Addition/Renovation
	Newtown Municipal Center	Newtown	OH	5,077	$168.59	Adaptive Reuse
	Nederland City Hall	Nederland	TX	4,983	$375.09	New
Correctional	County Sheriffs Office	Morgantown	WV	31,645	$297.28	New
	Nederland Public Safety Complex	Nederland	TX	21,189	$217.30	Adaptive Reuse
	Detention Center & Sheriffs Office	Spencer	IA	16,983	$421.41	New
	Ogle City Sheriff & Coroner Admin	Oregon	IL	15,377	$283.95	New
	Chautauqua City Jail & Sheriff	Sedan	KS	12,257	$299.47	New
Courthouse	Courthouse HVAC System Replacement	Gainesville	FL	101,000	$43.19	Renovation
	Courthouse Renovation & Restoration	Springfield	IL	47,720	$184.63	Renovation
Fire Department	College Station Fire Station #6	College Station	TX	25,133	$350.16	New
	Mt. Orab Fire Station	Village of Mt. Orab	OH	18,170	$180.68	New
	Richardson Fire Station No. 4	Richardson	TX	14,090	$411.12	New
	Willowfork Fire Station No. 2	Katy	TX	13,358	$302.19	New
	Joint Fire & Rescue Station	Newtown	OH	13,125	$194.79	Addition/Renovation
	Little Miami Fire & Rescue	Fairfax	OH	12,316	$231.30	New
	Fire Station No. 11	Fort Smith	AR	12,155	$350.36	New
	Pearisburg Fire Station	Pearisburg	VA	11,818	$225.50	New
	Ponderosa Fire Station No. 62	Spring	TX	11,163	$302.54	New
	El Dorado Hills Fire Station 84	El Dorado Hills	CA	10,869	$455.55	New
	Wayne Fire Department	Goshen	OH	10,000	$99.16	New
	Fire Station No. 40	Jacksonville	FL	9,703	$399.75	New
	Rosenberg Fire Station No. 3	Rosenberg	TX	8,479	$378.94	New
	Little Rock Fire Station No. 23	Little Rock	AR	8,291	$487.53	New
	Cleveland Volunteer Fire Station	Cleveland	MS	6,910	$340.81	New
Government	Council Center For Scouting	Fargo	ND	20,466	$204.40	New
	Camp Crook Ranger Station	Camp Crook	SD	4,880	$504.83	New
	Beaumont Municipal Tennis Center	Beaumont	TX	4,460	$369.39	Addition
	Florence Transit Hub	Florence	KY	3,115	$527.64	New
	Knox Area Rescue Ministries	Knoxville	TN	1,762	$694.90	New
	Entrance Station Lake Mead	Clark County	NV	480	$2,819.31	New
	Vehicle Charging Stations	Denton	TX	6 spaces	$7,233.53	New
Library	Dover Public Library	Dover	DE	46,424	$443.19	New
	Clinton-Macomb Public Library	Clinton Township	MI	24,723	$129.49	Adaptive Reuse
	Crozet Western Albemarle Library	Crozet	VA	23,199	$359.80	New
	Upper Tampa Bay Regional Library	Tampa	FL	13,630	$249.02	Addition/Renovation
	Palmetto Branch Library	Palmetto	GA	11,200	$496.97	New
	Regional Library Expansion	Valrico	FL	10,970	$305.27	Addition/Renovation
Miscellaneous	City of Pampa Animal Welfare	Pampa	TX	13,578	$289.19	New
	Royal Winnipeg Ballet Renovations	Manitoba	CA	13,237	$78.02	Renovation
	Senior Services - Kitchen Facility	Batavia	OH	6,000	$123.50	New
	Historical Site Locomotive Shelter	Bismarck	ND	2,100	$165.06	New
Office	Federal Building & Courthouse Modernization	Denver	CO	41,600	$357.90	Renovation
	Brazos County Tax Office	Bryan	TX	13,143	$307.93	New
	Illinois Water District Office	Lincoln	IL	8,974	$160.26	New
	Arkansas River Resource Center	Little Rock	AR	4,926	$588.40	New

2021 SQUARE FOOT TABLES

PROJECT	DESCRIPTION	CITY	STATE	SIZE	$/SF	NOTES
		Educational				
Athletic Facility	Physical Activity/Sports Science	Morgantown	WV	117,344	$244.33	New
	Indoor Football Practice Facility	Clemson	SC	81,992	$188.54	New
	Jesuit College Locker Room Addition	Dallas	TX	41,673	$225.42	Addition
	Multi-Purpose Gymnasium	Jacksonville	FL	22,844	$299.13	New
College Classroom	New Mexico Tech Geology Building	Socorro	NM	86,813	$325.86	New
	CSU Concourse & Training Room	Fort Collins	CO	53,050	$152.21	Addition/Renovation
	Jack Williamson Liberal Arts Center	Portales	NM	52,480	$244.58	Renovation
	UNLV Literature & Law Building	Las Vegas	NV	44,830	$203.35	Renovation
	Southern State Community College	Mount Orab	OH	43,833	$195.65	New
	SERT Building Iowa Lakes College	Estherville	IA	42,940	$134.09	Adaptive Reuse
Elementary	Cibolo Valley Elementary School	Cibolo	TX	153,130	$264.86	New
	Hill Farm Elementary School	Bryant	AR	88,800	$316.82	New
	CREC International Magnet School	South Windsor	CT	63,923	$435.46	New
	Pineville Elementary School	Pineville	WV	51,650	$231.89	New
	Crownpoint Elementary School	Crownpoint	NM	48,592	$432.07	New
	Janney Elementary School Addition	Washington	DC	10,000	$638.61	Addition
High School	Hmong College Prep Academy	Saint Paul	MN	154,434	$90.68	Addition/Renovation
	HFC High School North Building	Flossmoor	IL	136,555	$221.96	Addition/Renovation
	Somerset High School	Von Ormy	TX	125,800	$192.78	New
	Takoma Education Campus	Washington	DC	119,000	$238.43	Renovation
	High School Addition & Renovation	Decatur	GA	83,816	$276.83	Addition/Renovation
	Palmer Catholic Academy	Ponte Vedra Beach	FL	34,209	$150.00	New
	Elmwood High School Addition	Elmwood Park	IL	31,630	$371.42	Addition/Renovation
	High School Fine Arts Building	Heber Springs	AR	30,505	$433.80	New
	Alamo Heights High School Fine Arts	San Antonio	TX	25,536	$348.60	Addition/Renovation
	St. Patrick Catholic School	Jacksonville	FL	23,227	$319.96	New
	Springdale School Alteration	Corbett	OR	13,680	$136.51	Renovation
	Goddard School Addition Renovation	Anderson Township	OH	12,489	$90.96	Addition/Renovation
	ISD Outdoor Education Center	Sabine Pass	TX	10,193	$345.68	New
	Indian Mountain School Student Center	Lakeville	CT	9,335	$331.86	Addition
	First Impressions Academy	Fayetteville	NC	7,752	$190.94	New
	High School South Campus Field	Cincinnati	OH	5,580	$256.08	New
Middle School	Timberline Middle School	Waukee	IA	187,375	$171.88	New
	Red Bank Middle School	Chattanooga	TN	158,637	$284.58	New
	Jaime Escalante Middle School	Pharr	TX	156,538	$213.09	New
	Conservatory Green ECE-8 School	Denver	CO	113,616	$190.04	New
	Midland School Addition	Floral	AR	30,150	$255.54	Addition
Laboratory/Research	Science & Technology Building	Fayetteville	NC	65,048	$539.12	New
	NSU/US Geological Survey	Davie	FL	24,000	$112.03	Tenant Build-out
	Northeast Technology Center	Pryor	OK	11,909	$336.06	New
	Environmental Education Center	Bushkill Township	PA	9,275	$582.64	New
	Research & Education Center	Homestead	FL	5,760	$735.26	New
Multi-Purpose	BGSU Student Rec Center	Bowling Green	OH	179,549	$70.62	Renovation
	Kennedy Center Theatre/Studio Arts	Clinton	NY	96,100	$356.65	New
	Classroom & Administration Building	Houston	TX	65,234	$230.55	New
	NM State U Pete Domenici Building	Las Cruces	NM	53,341	$290.60	Addition/Renovation
	Widener University Freedom Hall	Chester	PA	36,700	$343.98	New
	Alumni Hall, Lincoln Park	Midland	PA	29,027	$280.51	New
	GSU Piedmont North Dining Hall	Atlanta	GA	12,300	$393.83	Addition
	NAU Dining Hall Expansion Phase II	Flagstaff	AZ	10,096	$502.26	New
	Neighborhood Resource Center	Richmond	TX	6,935	$216.82	New
	Heber Springs Cafeteria Remodel	Heber Springs	AR	5,585	$244.33	Renovation

All prices are updated to January 1, 2021 and are national averages.
For a more in-depth report of any of these buildings or additional case studies, contact
Design, Cost and Data at 800-533-5680, or go to www.DCD.com

SF-5

2021 SQUARE FOOT TABLES

PROJECT	DESCRIPTION	CITY	STATE	SIZE	$/SF	NOTES
				Hotels		
Hotels	Omni Dallas Hotel	Dallas	TX	1,161,450	$443.81	New
	John Ascuagas Nugget Hotel/Casino	Sparks	NV	449,820	$151.41	Addition
	Le Centre On Fourth Embassy Suites	Louisville	KY	408,229	$119.82	Adaptive Reuse
	Minneapolis Marriott West	Minneapolis	MN	237,362	$172.95	New
	Sheraton Centre Park Hotel	Dallas	TX	231,031	$247.51	New
	AmeriSuites	Chicago	IL	191,600	$156.50	Addition/Renovation
	Hampton Inn & Suites Hotel	Chicago	IL	162,000	$170.91	New
	Sheraton Harbor Island Hotel Tower	San Diego	CA	144,126	$208.37	Addition
	Compri Hotel	Los Angeles	CA	110,150	$184.54	New
	Best Western Columbia Hotel	San Diego	CA	108,040	$142.76	New
	Spooky Nook Warehouse Hotel	Manheim	PA	92,726	$135.95	Adaptive Reuse
	The Atrium Motel	Norfolk	VA	75,889	$145.99	New
	Staybridge Hotel At Preston Ridge	Alpharetta	GA	74,607	$187.19	New
	Hampton Inn & Suites	Allentown	PA	71,686	$156.12	New
	Hampton Inn Hotel	Carol Stream	IL	71,000	$191.72	New
	The Inn On Lake Superior	Duluth	MN	65,345	$154.22	New
	The Lancaster Hotel	Houston	TX	64,310	$322.73	Renovation
	Fairfield Inn	Helena	MT	31,009	$150.71	New
	The Edison Hotel	Miami	FL	28,875	$124.94	Renovation
	Western Executive Inn	Billings	MT	21,984	$111.55	New
	Country Hearth Inn	Preston	MN	21,028	$130.97	New
	Lawrence Welk Resort Hotel	Escondido	CA	19,874	$149.39	Addition
	Hanalei Hotel Conference Center	San Diego	CA	8,587	$237.22	Addition
	Summit At Vail, Multi-Purpose Lodge	Vail	CO	6,000	$283.23	New

All prices are updated to January 1, 2021 and are national averages.
For a more in-depth report of any of these buildings or additional case studies, contact
Design, Cost and Data at 800-533-5680, or go to www.DCD.com

2021 SQUARE FOOT TABLES

PROJECT	DESCRIPTION	CITY	STATE	SIZE	$/SF	NOTES

Industrial

PROJECT	DESCRIPTION	CITY	STATE	SIZE	$/SF	NOTES
Manufacturing	Brentwood Industries Manufacturing	Reading	PA	205,000	$43.98	New
	Lee Steel Corporate Plant	Romulus	MI	200,625	$95.66	New
	Siemens Westinghouse Fuel Cell Facility	Munhall	PA	191,090	$104.16	New
	Manufacturing Plant & Headquarters	Lansing	MI	188,975	$100.99	Addition
	Concepts Direct	Longmont	CO	117,900	$123.18	New
	Nypro Inc.	Clinton	MA	102,475	$126.36	Addition
	SWF Industrial	Wrightsville	PA	76,218	$77.85	New
	Headquarters & Manufacturing Facility	Lower Nazareth Township	PA	62,980	$37.29	Renovation
	Aerzen USA (Office/Manufacturing)	Coatesville	PA	40,000	$190.27	New
	Lee Steel Corporate Expansion	Wyoming	MI	34,821	$80.83	New
	Prescott Aerospace	Prescott	AZ	31,400	$116.45	New
	Battery Innovation Center	Newberry	IN	30,080	$454.35	New
	American Steel	Billings	MT	25,957	$84.46	New
	Phillip S. Luttazi Town Garage	Dover	MA	21,913	$145.33	New
	ITT Flygt - Industrial Facility	Milford	OH	16,991	$137.81	New
	Broadmoor Golf Maintenance	Colorado Springs	CO	16,064	$249.82	New
	Cooper B-Line Expansion	Highland	IL	15,290	$176.51	Addition/Renovation
	Robberson Ford Collision Center	Bend	OR	15,089	$144.89	New
	Brown Industrial Building	Truckee	CA	13,345	$143.13	New
	CTC Vehicle Maintenance Shops	Killeen	TX	11,250	$135.34	New
	Storage & Shop Facility	Billings	MT	8,763	$93.40	New
	Central Plant with Equipment Bay	Mesa	AZ	8,500	$962.99	New
Office	Woodlands Business Center	Richmond	VA	48,000	$83.66	New
	Wiregrass Research Center	Headland	AL	9,740	$258.37	New
Office/Warehouse	American Superconductor	Devens	MA	354,000	$177.79	New
	Castcon Stone Inc.	Saxonburg	PA	47,000	$119.08	New
	Minnesota DNR Headquarters	Tower	MN	37,802	$176.35	New
	Office & Warehouse	Miami	FL	14,815	$144.48	New
	DOT Office & Maintenance Building	Hillsboro	OH	10,876	$254.87	New
Warehouse	Distribution Center	Windsor	CT	303,750	$36.25	New
	Zany Brainy Distribution Center	Bridgeport	NJ	250,000	$43.29	New
	Galderma - Warehouse	Fort Worth	TX	70,000	$93.26	New
	Manzana Products Warehouse	Sebastopol	CA	41,395	$69.46	New
	Tactical Equip Maintenance Facility	Fort Campbell	KY	35,290	$250.61	New
	Sonoma Wine Company Canopy	Graton	CA	26,000	$58.96	New
	Administration/Chemical Storage Building	Killeen	TX	23,837	$356.50	New
	DOT Truck Storage Building	Hillsboro	OH	18,400	$97.26	New
	F.I. Storage Facility	Kentwood	MI	13,125	$71.52	New
	50 Columbia Drive Warehouse	Pooler	GA	10,000	$89.76	New
	DOT Salt Storage Building	Hillsboro	OH	9,100	$98.06	New
	Organizational Storage Facility	Fort Campbell	KY	8,040	$148.79	New
	Dwan Maintenance Building	Bloomington	MN	7,240	$200.94	Addition/Renovation
	Maintenance/Storage Building	Batavia	OH	7,200	$86.43	New
	DOT Cold Storage Building	Hillsboro	OH	5,040	$104.56	New
	Job Corp Warehouse	Hartford	CT	3,800	$495.71	New
	DOT Materials Storage Building	Hillsboro	OH	1,920	$116.90	New
	Aerial Vehicle Storage	Fort Campbell	KY	1,800	$223.41	New
	Petro, Oil, Lubricant Storage	Fort Campbell	KY	640	$312.31	New
	Hazardous Waste Storage Building	Fort Campbell	KY	640	$333.76	New

All prices are updated to January 1, 2021 and are national averages.
For a more in-depth report of any of these buildings or additional case studies, contact
Design, Cost and Data at 800-533-5680, or go to www.DCD.com

SF-7

2021 SQUARE FOOT TABLES

PROJECT	DESCRIPTION	CITY	STATE	SIZE	$/SF	NOTES
		Medical				
Clinic	HealthCare Emergency/Trauma Center	Topeka	KS	115,000	$421.22	Addition
	Sadler Clinic (Shell Only)	Conroe	TX	61,599	$131.13	New
	Pinellas County Health Department	Largo	FL	54,965	$273.25	Retrofit
	Sanford Moorhead Clinic	Moorhead	MN	49,250	$267.79	New
	County Health Department	Port Charlotte	FL	47,564	$303.63	New
	Sadler Clinic	Conroe	TX	41,066	$124.23	Tenant Build-out
	PineMed Medical Plaza	The Woodlands	TX	30,398	$138.54	New
	Outpatient Specialty Clinic	Vancouver	WA	20,139	$344.70	New
	Ambulatory Surgery Center	Stroudsburg	PA	19,929	$369.95	Addition/Renovation
	Thundermist Health Center	West Warwick	RI	18,217	$213.31	Adaptive Reuse
	E Texas Community Health Services	Nacogdoches	TX	12,500	$101.02	Retrofit
	North Mobile Health Center	Mt. Vernon	AL	6,765	$301.35	New
Dental Office	Construct Dental Clinic Roseburg	Roseburg	OR	7,750	$524.39	New
	Kitchens Pediatric Dental Clinic	Little Rock	AR	6,068	$315.67	New
	Dental Office Shell & Parking	Olympia	WA	5,302	$179.11	New
	Evans Family Dental	Austin	TX	2,354	$219.51	Tenant Build-out
Hospital	Union Hospital Addition	Terre Haute	IN	492,348	$329.92	Addition
	Regional Medical Center	Lafayette	LA	410,273	$552.16	New
	Houston Medical Pavilion	Warner Robins	GA	180,000	$65.05	Adaptive Reuse
	Langley AFB Hospital Renovation	Langley Air Force Base	VA	160,000	$551.26	Renovation
	Cass Regional Medical Center	Harrisonville	MO	137,524	$387.62	New
	Texas Spine & Joint Hospital	Tyler	TX	115,789	$256.73	Addition/Renovation
	Oktibbeha County Hospital Expansion	Starkville	MS	87,116	$336.41	New
	Childrens Mercy Hospital	Independence	MO	54,682	$324.12	New
	Oktibbeha County Hospital Renovation	Starkville	MS	30,263	$167.91	Renovation
	El Rio Community Health Center	Tucson	AZ	26,998	$247.56	New
	Pondella Public Health Center	Fort Myers	FL	26,400	$308.35	Renovation
	Topeka Ear Nose & Throat	Topeka	KS	24,073	$319.70	New
	Rapha Primary Care	Fayetteville	NC	19,907	$137.25	Renovation
	UNM Hospitals North Valley Center	Albuquerque	NM	16,500	$323.25	New
	El Rio Community Health Center	Tucson	AZ	14,000	$281.06	New
	VA Medical Center Area G Renovation	Houston	TX	12,000	$341.99	Renovation
	Surgical Suite Expansion	Dobbs Ferry	NY	9,000	$390.34	Renovation
	Legacy Emergency Room	Allen	TX	8,432	$647.41	New
	Oral & Maxillofacial Surgery Center	Fayetteville	NC	2,214	$550.86	Renovation
Nursing Home/Rehab	Senior Living Community	Hoschton	GA	56,251	$143.56	New
	Retirement Community	Carlisle	PA	47,075	$161.66	Addition/Renovation
	Assisted Living & Memory Center	Dacula	GA	38,221	$166.98	New
	Homestead Village Nursing Care	Lancaster	PA	28,149	$109.33	Renovation
	Jewish Services For The Aging	Tucson	AZ	24,993	$203.54	New
	Short-Term Rehabilitation	Olathe	KS	13,800	$273.81	Addition
	St. Katharine Retirement Center	El Reno	OK	12,000	$323.60	New/Addition
Office	Orthopedic Hospital/Medical Office	Allentown	PA	79,807	$205.10	Renovation
	Tomball Medical Office Building	Tomball	TX	54,380	$139.46	New
	Olathe Health Education Center	Olathe	KS	50,258	$311.86	New
	Home & Hospice Care	Providence	RI	47,734	$196.72	Renovation
	NE Georgia Medical Plaza 400	Dawsonville	GA	26,997	$274.47	Adaptive Reuse
	VA Medical Center/Pharmacy	Waco	TX	19,171	$159.81	Renovation
	Cancer Specialists of North Florida	Jacksonville	FL	18,654	$291.51	New
	MJHS Hospice Residence	N.Y.C.	NY	12,500	$150.84	Renovation
	Medical Office Building	Pelham	NH	8,399	$271.46	New
	Podiatry Group	Marietta	GA	6,768	$55.20	Renovation
	Marietta Podiatry Group	Marietta	GA	4,400	$252.11	New

All prices are updated to January 1, 2021 and are national averages.
For a more in-depth report of any of these buildings or additional case studies, contact
Design, Cost and Data at 800-533-5680, or go to www.DCD.com

2021 SQUARE FOOT TABLES

PROJECT	DESCRIPTION	CITY	STATE	SIZE	$/SF	NOTES
			Office			
Office	Restaurant Support Center	Lenexa	KS	186,465	$266.92	New
	5000 NASA Boulevard	Fairmont	WY	132,000	$284.61	New
	Rockford Construction Office	Grand Rapids	MI	71,144	$113.72	Adaptive Reuse
	Woodlawn Office Bldg. (Shell Only)	Louisville	KY	60,000	$143.86	New
	Rosecrance Ware Center	Rockford	IL	44,800	$141.98	Adaptive Reuse
	Professional Center (Shell)	White Marsh	MD	43,025	$180.98	New
	Swan Skyline Office Plaza (Shell)	Tucson	AZ	37,200	$122.07	New
	Infinite Energy Phase IV	Gainesville	FL	36,500	$319.97	New
	Freedom Plaza Building	Cookeville	TN	28,488	$250.17	New
	Landmark Professional Building	Clayton	NC	27,231	$237.21	New
	FC Gulf Freeway Building (Shell)	Houston	TX	24,084	$270.05	New
	White Street Building (Shell Only)	Marietta	GA	23,809	$220.38	New
	Columbia Shores Office Condo	Vancouver	WA	22,574	$182.90	New
	Office & Design Studio	Chicago	IL	20,244	$128.41	Tenant Build-out
	Pinnacle III Office Tenant Finish	Leawood	KS	18,409	$61.13	Tenant Build-out
	Commerce Park (Shell)	Suwanee	GA	17,097	$125.37	New
	PCWA Business Center Interior	Auburn	CA	12,085	$63.59	Renovation
	Tenth Avenue Holdings Offices	N.Y.C.	NY	11,000	$89.26	Tenant Build-out
	Office Park - Building A (Shell)	Fort Collins	CO	10,000	$274.11	New
	Longshoremens Welfare Fund Building	Savannah	GA	8,160	$402.45	New
	Garry Street Office Building	Manitoba	CA	7,506	$160.58	Renovation
	Reserve Advisors	Milwaukee	WI	5,300	$51.48	Tenant Build-out
	510 Armory Street Office	Boston	MA	5,100	$98.57	Renovation
	Offices of Bonsall Shafferman	Bethlehem	PA	4,950	$72.91	Tenant Build-out
	FCT Capital Partners	Houston	TX	4,100	$85.29	Tenant Build-out
	Martin Rogers Associates Office	Wilkes-Barre	PA	4,000	$136.48	Addition/Renovation
	Cowan & Kohne Financial	Suwanee	GA	3,713	$128.52	Tenant Build-out
	212 Archer Street Office	Bel Air	MD	3,600	$181.80	New
	Utilities Analyses Inc.	Suwanee	GA	3,513	$119.21	Tenant Build-out
	Visual Lizard Interior Fit-Up	Manitoba	CA	2,430	$95.65	Tenant Build-out
	Richardson State Farm	Houston	TX	2,200	$115.65	Tenant Build-out
Mixed-Use	Office/Retail/Parking Mixed-Use	Jackson	MS	228,407	$262.41	New
	Korte & Luitjohan Office & Shop	Highland	IL	26,000	$126.09	New
Medical Office	Evanston Medical Office Building	Evanston	WY	9,157	$289.02	New
	Advanced Medical Group	Suwanee	GA	4,433	$122.72	Tenant Build-out
	Bothell Dental Office Build Out	Bothell	WA	2,203	$203.79	Tenant Build-out
Headquarters	CONSUL Energy Corporation Headquarters	Southpointe, Canonsburg	PA	317,500	$243.40	New
	Fairmont Supply Corporate Headquarters	Southpointe, Canonsburg	PA	75,255	$160.85	New
	Practice Velocity Corporate Headquarters	Machesney Park	IL	64,318	$101.07	Adaptive Reuse
	Enterprise Integration Headquarters	Jacksonville	FL	57,723	$62.88	Renovation
	Linear Technology	Cary	NC	20,000	$326.65	New
	PIPS Technology Inc.	Knoxville	TN	19,884	$259.65	New
	Lee Steel Corporate Offices	Novi	MI	15,781	$172.57	Renovation
	Weaver Cooke Headquarters	Goldsboro	NC	15,464	$310.53	New
	In Capital Holdings	Boca Raton	FL	13,000	$174.52	Tenant Build-out
	ACCION Regional Headquarters	Albuquerque	NM	7,580	$334.09	New
Civic Office	Miss Department of Environmental Quality	Jackson	MS	121,170	$97.32	Renovation
	County Central Office Complex	Pensacola	FL	74,630	$294.96	New
	State of WV Office Building	Fairmont	WV	70,442	$262.81	New
	JAX Chamber of Commerce Renovation	Jacksonville	FL	20,110	$216.99	Renovation

All prices are updated to January 1, 2021 and are national averages.
For a more in-depth report of any of these buildings or additional case studies, contact
Design, Cost and Data at 800-533-5680, or go to www.DCD.com

SF-9

2021 SQUARE FOOT TABLES

PROJECT	DESCRIPTION	CITY	STATE	SIZE	$/SF	NOTES
Recreational						
Educational	The Pavilion at Ole Miss	Oxford	MS	235,301	$461.96	New
	University Laker Turf Building	Allendale	MI	137,662	$153.17	New
	CSU Recreation Center	Chico	CA	110,245	$476.63	New
	Intramural Recreation Penn State	State College	PA	59,303	$422.80	Addition/Renovation
	Center For Women's Athletics	Fayetteville	AR	39,183	$376.49	New
	Pickens Recreation Center	Pickens	SC	20,400	$187.94	New
	High School Concessions & Press Box	Loganville	GA	1,506	$482.87	New
Health Club	Brooklyn Yard Fitness Club (Shell)	Portland	OR	63,987	$118.32	New
	Title Boxing Club	Cedar Hill	TX	4,330	$73.35	Tenant Build-out
Recreational	Community Recreation Center	Williston	ND	223,787	$459.86	New
	Spirit Lake Casino & Resort	St. Michael	ND	112,277	$62.37	Renovation
	Phipps Tropical Forest	Pittsburgh	PA	80,000	$335.63	New
	Family Recreation Center	Colonie	NY	70,256	$245.30	New
	Youth Activity Center	Joplin	MO	62,056	$103.88	New
	New Holland Recreational Center	New Holland	PA	51,256	$135.45	Renovation
	Community College Recreation Center	Cedar Rapids	IA	43,500	$170.93	New
	Anderson Recreation Center	Anderson	SC	34,282	$383.25	New
	East Park Community Center	Nashville	TN	33,000	$307.43	New
	Church Family Life Center	Clemson	SC	31,509	$331.75	New/Renovation
	C.K. Ray Recreation Center	Conroe	TX	30,380	$156.57	Addition/Renovation
	St. Raphael Athletic & Wellness Center	Pawtucket	RI	30,268	$271.07	New
	The Forge For Families	Houston	TX	29,860	$255.57	New
	Christian Life Center	Birmingham	MI	26,966	$349.81	Addition
	Children's Sports Center	Woodbury	MN	26,219	$125.10	New
	Trinity River Audubon Center	Dallas	TX	20,791	$964.61	New
	Baptist Church Activity Center	Indianapolis	IN	16,636	$154.67	New
	Boys & Girls Club Syracuse	Syracuse	NY	12,107	$189.04	Addition
	Job Corp Recreational Building	Hartford	CT	11,300	$212.11	New
	Presbyterian Family Life Center	Strawberry Plains	TN	11,236	$172.35	New
	McDaniel Yacht Basin	North East	MD	7,620	$194.11	New
	Bicentennial Park	Cincinnati	OH	4,050	$909.31	New
	Bahosky Softball Complex	Bronx	NY	3,800	$491.75	New
Swimming Center	Resort & Indoor Waterpark	Cortland	NY	175,060	$274.72	New
	The Aquatic Center	Tunica	MS	45,008	$327.33	New
	Spirit Lake Phase 4	St. Michael	ND	26,630	$374.25	Addition/Renovation
	Family Aquatic Center	Beachwood	OH	7,500	$1,216.24	New
	Community Aquatic Park & Center	Billings	MT	6,730	$764.93	New
Theater	Cinema & IMAX Theatre	Lansing	MI	13,750	$231.22	New
	Academy Theater	N.Y.C.	NY	4,593	$234.91	Renovation
YMCA	David D. Hunting YMCA	Grand Rapids	MI	162,966	$210.71	New
	YMCA Recreational Center	Ann Arbor	MI	83,377	$286.80	New
	Floyd Co. YMCA & Aquatic Center	New Albany	IN	82,324	$349.20	New
	Wade Walker Park Family YMCA	Stone Mountain	GA	59,134	$364.48	New
	Alexandria YMCA	Alexandria	MN	55,150	$176.35	New
	Greater Nashua YMCA	Nashua	NH	49,980	$217.06	New
	Lancaster YMCA Harrisburg Ave.	Lancaster	PA	42,502	$369.13	New
	Greater Kingsport Family YMCA	Kingsport	TN	40,007	$263.67	New
	Eastside YMCA	Knoxville	TN	39,984	$261.97	New
	Highland County Family YMCA	Hillsboro	OH	33,228	$156.09	New
	Houston Texans YMCA	Houston	TX	31,628	$370.40	New
	Cypress Creek YMCA	Houston	TX	25,699	$156.22	Addition/Renovation

All prices are updated to January 1, 2021 and are national averages.
For a more in-depth report of any of these buildings or additional case studies, contact
Design, Cost and Data at 800-533-5680, or go to www.DCD.com

2021 SQUARE FOOT TABLES

PROJECT	DESCRIPTION	CITY	STATE	SIZE	$/SF	NOTES

Religious

PROJECT	DESCRIPTION	CITY	STATE	SIZE	$/SF	NOTES
Church	First United Methodist Church	Orlando	FL	121,536	$291.57	New
	Beautiful Savior Lutheran Church	Plymouth	MN	69,700	$119.25	New
	Solid Rock Baptist Church	Berlin	NJ	58,359	$107.82	New
	North Side Baptist Church	Greenville	SC	48,087	$275.35	New
	St. Martha Catholic Church	Porter	TX	46,748	$544.05	New
	Gracepoint Gospel Fellowship Church	Ramapo	NY	46,595	$202.36	New
	Good Shepherd Methodist Church	Odessa	TX	41,003	$322.30	New
	Davisville Church Addition	Southampton	PA	36,090	$190.55	Addition
	Immaculate Catholic Church	Columbia	IL	34,000	$266.45	New
	Keystone Community Church	Ada	MI	29,775	$156.89	New
	Grace Church	Des Moines	IA	29,296	$212.86	New
	Good Shepherd Church	Naperville	IL	27,869	$210.85	Addition/Renovation
	River Hills Baptist Church	Corpus Christi	TX	27,404	$319.22	New
	Good Shepherd Catholic Church	Smithville	MO	24,810	$219.30	New
	Prince of Peace Catholic Church	Chesapeake	VA	24,740	$274.77	Addition/Renovation
	Sanctuary Addition Christian Church	Oklahoma City	OK	23,820	$164.70	Addition
	St. Sylvester Catholic Church	Gulf Breeze	FL	22,000	$431.86	New
	St. Peters Catholic Sanctuary	Fallbrook	CA	20,764	$475.86	New
	Notre Dame Catholic Church	Houston	TX	20,280	$438.99	New
	St Eugene Catholic Church	Oklahoma City	OK	20,000	$481.16	New
	Chapin Presbyterian Church	Chapin	SC	19,900	$350.82	New
	St. Michaels Catholic Church	Glen Allen	VA	19,770	$376.38	New
	Shrine of Holy Spirit	Branson	MO	19,200	$358.18	New
	Wildwood United Methodist Church	Magnolia	TX	19,000	$255.93	New
	Episcopal Church of the Nativity	Scottsdale	AZ	18,288	$128.00	Adaptive Reuse
	Hardin Church of Christ	Knoxville	TN	17,149	$128.68	New
	First United Methodist Church	Crossville	TN	15,816	$503.64	New
	Good Shepherd Episcopal Church	Silver Spring	MD	15,200	$300.51	Addition/Renovation
	Ascension Catholic Church	LaPlace	LA	15,057	$396.69	New
	St. Patrick Catholic Church	Jacksonville	FL	14,139	$283.55	New
	Our Lady of Guadalupe Catholic Church	Rosenberg	TX	12,910	$437.35	New
	St. Timothy's Episcopal Church	Creve Coeur	MO	12,682	$218.03	New/Renovation
	St. Paul Lutheran Church	Pomaria	SC	12,072	$356.98	New
	Covenant Baptist Church	Florida City	FL	10,725	$266.36	New
	First United Methodist Church	Katy	TX	10,503	$323.51	Addition/Renovation
	Lake Ann United Methodist Church	Lake Ann	MI	9,975	$223.55	New
	United Methodist Church	Odenton	MD	8,783	$349.29	New
	Kent R. Hance Chapel at Texas Tech	Lubbock	TX	6,530	$634.03	New
	Haven for Hope Chapel	San Antonio	TX	2,232	$476.23	New
Multi-Purpose	Baptist Church Multi-Purpose Bldg	Maryville	TN	41,656	$118.82	New
	Good Shepherd Parish Center	San Diego	CA	28,752	$163.89	New
	Christian Life Center	Kansas City	MO	26,320	$318.31	New
	Baptist Church Outreach Center	Fort Smith	AR	25,000	$279.49	New
	St Rafael Administration Building	San Diego	CA	24,276	$176.52	New
	Catholic Church Social Hall	Chula Vista	CA	23,596	$293.69	New
	United Methodist Church	West Chester	PA	11,935	$256.82	Addition/Renovation
	Student Ministry Center	Knoxville	TN	11,700	$318.18	New
	Catholic Pastoral Ministries Center	Spring	TX	10,135	$408.50	New
	Christian Renewal Center	Dickinson	TX	8,500	$224.58	New
	Holy Family St Lawrence Parish Center	Essex Junction	VT	7,900	$235.52	New
	Presbyterian Church Addition	Gap	PA	7,414	$270.36	Addition/Renovation
	New Hope Church Addition/Alteration	Saint Louis	MO	5,564	$207.01	Renovation

All prices are updated to January 1, 2021 and are national averages.
For a more in-depth report of any of these buildings or additional case studies, contact
Design, Cost and Data at 800-533-5680, or go to www.DCD.com

SF-11

2021 SQUARE FOOT TABLES

PROJECT	DESCRIPTION	CITY	STATE	SIZE	$/SF	NOTES
			Residential			
Apartment	Solace Apartments	Virginia Beach	VA	331,681	$104.79	New
	1221 Broadway Lofts	San Antonio	TX	205,137	$148.00	Adaptive Reuse
	Sustainable Fellwood Phase I	Savannah	GA	124,037	$138.45	New
	Kelly Cullen Community	San Francisco	CA	98,385	$573.66	Adaptive Reuse
	Bachelors Enlisted Quarters	Camp Williams	UT	76,253	$260.84	New
	Mockingbird Terrace Homes	Louisville	KY	71,110	$165.78	New
	Homeless Men's Residential	San Antonio	TX	67,908	$274.80	New
	Homeless Women's/Family Residence	San Antonio	TX	60,182	$264.80	New
	Magnolia Place	Lancaster	PA	39,714	$166.99	New
	Elkins First Ward Apartments	Elkins	WV	27,000	$119.88	Adaptive Reuse
	Young Burlington Apartments	Los Angeles	CA	24,399	$214.84	New
	The Lofts at 300 Bowman	Dickson City	PA	23,900	$89.56	Adaptive Reuse
	Peaceful Paths Emergency Svc Campus	Gainesville	FL	22,535	$153.73	New
	Wylie House - Ronald McDonald House	Kansas City	MO	21,885	$194.99	New
	Dogwood Manor Apartments	Oak Ridge	TN	19,975	$176.85	New
	Anderson Village Multi-Family	Austin	TX	12,500	$295.24	New
	Salvation Army Sally's House	Houston	TX	7,812	$257.65	Addition
	Stones River Apartment Complex	Murfreesboro	TN	7,548	$238.26	Addition
	Sunshine Park Apartments Renovation	Gainesville	FL	2,252	$126.29	Renovation
Assisted Living	Kenmore Apartments Senior Housing	Chicago	IL	90,528	$220.91	Renovation
	Country Meadows Retirement	Allentown	PA	53,237	$169.03	New
	Creekside Village Assisted Living	Harrisburg	PA	16,150	$146.90	New
	Landis Homes Retirement Community	Lititz	PA	14,255	$69.43	Renovation
Dormitory	NSU Graduate Student Housing	Davie	FL	203,500	$226.96	Renovation
	Rider University Student Housing	Lawrenceville	NJ	50,500	$244.12	New
	JWU Biscayne Commons Dormitory	Miami	FL	40,048	$267.78	New
	College Residence Dorm	Bloomfield	NJ	25,980	$329.72	Renovation
Single-Family Home	Island Residence	Grosse Ile	MI	19,237	$711.73	New
	MG Residence Restoration	Williamston	MI	9,768	$64.01	Renovation
	Concepcion House	Coral Gables	FL	6,067	$367.06	New
	Leal House	Miami	FL	5,935	$253.59	New/Renovation
	Monserrate Street Residence	Coral Gables	FL	5,885	$458.63	New
	Private Residence	Newburgh	IN	5,566	$365.63	New
	Private Residence	Austin	MN	5,489	$150.64	New
	Private Residence	Lake Wallenpaupack	PA	4,845	$329.92	New
	Island in the Grove	Boca Raton	FL	4,701	$377.14	New
	Private Residence	Benson	AZ	3,660	$183.96	New
	Fairhope Green Home	Fairhope	AL	3,610	$209.44	New
	PATH Concept House	Omaha	NE	3,490	$80.23	New
	Private Residence	La Jolla	CA	3,420	$370.20	New
	Solar House - Private Residence	Fly Creek	NY	3,304	$185.27	New
	Elliott Residence	Fort Collins	CO	3,300	$279.38	New
	Renfrew House	Manitoba	CA	3,206	$199.94	New
	Rosado I Hansen Residence	Tucson	AZ	3,175	$139.47	New
	306 W. Waldburg Residence	Savannah	GA	2,588	$174.90	New
	Nutter Green Home	Milford	OH	2,289	$166.78	New
	Guest House Residence	Ahwatuckee	AZ	1,913	$407.10	New
	Kiwi House	Baton Rouge	LA	1,515	$162.58	New
	Private Residence Renovation	Shavertown	PA	810	$168.99	Renovation

UNITED STATES INFLATION RATES
July 1, 1970 to June 30, 2020, AC&E PUBLISHING CO. INDEX

Start July 1

YEAR	AVERAGE %	LABOR %	MATERIAL %	CUMULATIVE %
1970-71	8	8	5	0
1971-72	11	15	9	11
1972-73	7	5	10	18
1973-74	13	9	17	31
1974-75	13	10	16	44
1975-76	8	7	15	52
1976-77	10	5	15	62
1977-78	10	5	13	72
1978-79	7	6	8	79
1979-80	10	5	13	89
1980-81	10	11	9	99
1981-82	10	11	10	109
1982-83	5	9	3	114
1983-84	4	5	3	118
1984-85	3	3	3	121
1985-86	3	2	3	124
1986-87	3	3	3	127
1987-88	3	3	4	130
1988-89	4	3	4	134
1989-90	3	3	3	137
1990-91	3	3	4	140
1991-92	4	4	4	144
1992-93	4	3	5	148
1993-94	4	3	5	152
1994-95	4	3	6	156
1995-96	4	3	6	160
1996-97	3	3	3	163
1997-98	4	3	4	167
1998-99	3	3	2	170
1999-2000	3	2	3	173
2000-2001	4	4	4	177
2001-2002	3	4	3	180
2002-2003	4	5	3	184
2003-2004	3	5	2	187
2004-2005	8	4	10	195
2005-2006	8	6	9	203
2006-2007	8	4	11	212
2007-2008	3	4	3	215
2008-2009	4	3	5	219
2009-2010	3	5	1	222
2010-2011	2	3	2	224
2011-2012	3	3	3	227
2012-2013	3	3.5	3	230
2013-2014	2	1.8	2	232
2014-2015	1.4	1.8	1	233.4
2015-2016	1	1.8	0	234.4
2016-2017	1	1.8	0	235.4
2017-2018	5	4.5	5	240.4
2018-2019	3.5	2.3	4.6	244
2019-2020	2	1.5	2.5	246
2020-2021	2.5	2	3	248.5

Design Cost Data™ **DCD**

INDEX

Other Estimating References from BNi Building News

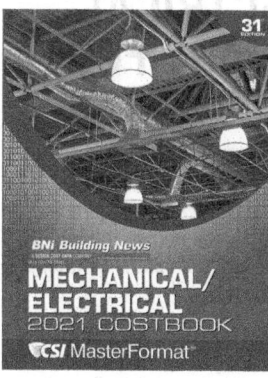

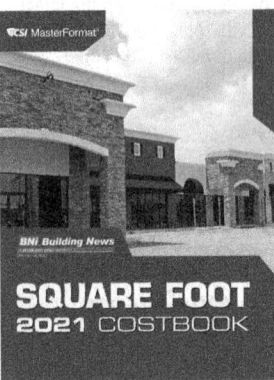

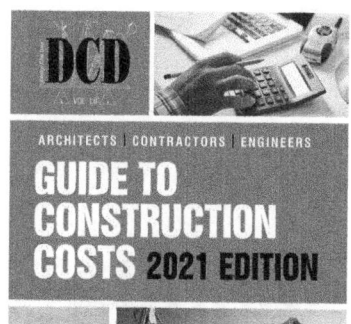

Construction Project Log Book

The 365 Daily Work Log pages let you keep a detailed record for each day of the year. Additional forms such as Accident Reports and numerous checklists help make sure that you're covered.

With *Construction Project Log Book* those tasks are made as simple as possible.

Document every shipment, machinery rental, delivery, delay, and weather condition — all items that can affect productivity. With the interactive forms you can keep this information on your computer and enter new data daily into the interactive PDF forms. There are forms that actually do the math for you — eliminating typical mistakes.

$39.95

Maintenance Manager's Standard Manual 6th Edition

Since it was first published in 1993, the *Maintenance Manager's Standard Manual* has indeed become the STANDARD reference in the field.

This brand-new Sixth Edition brings it completely up to date, incorporating the latest technology and best practices in all aspects of maintenance management.

Whether you are a facilities manager, engineer, property owner, developer, or anyone else responsible for maintenance operations; not only does it give you all of the essential ingredients for understanding and carrying out successful day-to-day management of maintenance activities, it provides you with an integrated plan for continuous improvement of the maintenance function.

$99.95

100% Satisfaction Guaranteed!

If not completely satisfied, return the item within 30 days for a complete refund of the purchase price.

This has been the BNi policy for over 74 years.

Is there a title you need and can't find?

Give us a call at 1-888-264-2665; we'll be glad to help.

Order online: www.bnibooks.com

Find hundreds of construction references, forms and contracts to help you with your construction business.

NOTES